SYSTEMS BIOLOGY

Systems Biology

International Research and Development

by

MARVIN CASSMAN
San Francisco, CA, U.S.A.

ADAM ARKIN
University of California, Berkeley, CA, U.S.A.

FRANK DOYLE
University of California, Santa Barbara, CA, U.S.A.

FUMIAKI KATAGIRI
University of Minnesota, St. Paul, MN, U.S.A.

DOUGLAS LAUFFENBURGER
Massachusetts Institute of Technology, Cambridge, MA, U.S.A.

and

CYNTHIA STOKES
Entelos Inc, Foster City, CA, U.S.A.

Springer

A C.I.P. Catalogue record for this book is available from the Library of Congress.

ISBN-10 1-4020-5467-X (HB)
ISBN-13 978-1-4020-5467-9 (HB)
ISBN-10 1-4020-5468-8 (e-book)
ISBN-13 978-1-4020-5468-6 (e-book)

Published by Springer,
P.O. Box 17, 3300 AA Dordrecht, The Netherlands.

www.springer.com

This document was sponsored by the National Science Foundation (NSF) and other agencies of the U.S. Government under awards from the NSF (ENG-0104476) and the Army Research Office (DAAD19-03-1-0067) awarded to the World Technology Evaluation Center, Inc. The government has certain rights in this material. Any opinions, findings, and conclusions or recommendations expressed in this material are those of the authors and do not necessarily reflect the views of the United States Government, the authors' parent institutions, or WTEC, Inc.

Printed on acid-free paper

WTEC Panel on Systems Biology

Marvin Cassman (chair)
Retired; San Francisco, U.S.A.
Adam Arkin
University of California, Berkeley, CA, U.S.A.
Frank Doyle
University of California, Santa Barbara, CA, U.S.A.
Fumiaki Katagiri
University of Minnesota, St. Paul, MN, U.S.A.
Douglas Lauffenburger
Massachusetts Institute of Technology, Cambridge, MA, U.S.A.
Cynthia Stokes
Entelos, Inc., Foster City, CA, U.S.A.

Abstract

The current textbook image of biological processes is that of a static model of loosely linked, highly detailed, molecular devices. However, every biologist knows that dynamic processes drive biology. Systems biology is defined for the purpose of this study as the understanding of biological network behaviors, and in particular their dynamic aspects, which requires the utilization of mathematical modeling tightly linked to experiment. This involves a variety of approaches, such as the identification and validation of networks, the creation of appropriate datasets, the development of tools for data acquisition and software development, and the use of modeling and simulation software in close linkage with experiment. All of these are discussed in this report. Of course, the definition becomes ambiguous at the margins. But at the core is the focus on networks, which makes it clear that the goal is to understand the operation of the systems, rather than the component parts. The panel concluded that the U.S. is currently ahead of the rest of the world in systems biology, largely because of earlier investment over the past five to seven years by funding organizations and research institutions. This is reflected in a large number of active research groups, and educational programs, and a diverse and growing funding base. However, there is evidence of rapid development outside the U.S., much of it begun in the last two to three years. It must be stressed that the attempt to incorporate the details of molecular events obtained over the past half century into a dynamic picture of network behavior in biological systems is only just beginning, in the U.S. and elsewhere. In particular, progress in the core activity of systems biology—modeling tied to experiment—is still limited. Progress would be facilitated by strong international collaborations in training, research, and infrastructure. Overall, however, the picture is of an active field in the early stages of explosive growth.

WTEC Mission

WTEC provides assessments of foreign research and development in selected technologies under awards from the National Science Foundation, the Office of Naval Research, and other agencies. Formerly part of Loyola College's International Technology Research Institute, WTEC is now a separate nonprofit research institute. Michael Reischman, Deputy Assistant Director for Engineering, is NSF Program Director for WTEC. Sponsors interested in international technology assessments and related studies can provide support for the program through NSF or directly through separate grants to WTEC.

WTEC's mission is to inform U.S. scientists, engineers, and policymakers of global trends in science and technology. WTEC assessments cover basic research, advanced development, and applications. Panels of typically six technical experts conduct WTEC assessments. Panelists are leading authorities in their field, technically active, and knowledgeable about U.S. and foreign research programs. As part of the assessment process, panels visit and carry out extensive discussions with foreign scientists and engineers in their labs.

The WTEC staff helps select topics, recruits expert panelists, arranges study visits to foreign laboratories, organizes workshop presentations, and finally, edits and disseminates the final reports.

Executive Editor: Dr. Marvin Cassman, Former Director, National Institute of General Medical Sciences, NIH

Series Editor: Dr. R.D. Shelton, President WTEC

World Technology Evaluation Center, Inc. (WTEC)

Foreword

> We have come to know that our ability to survive and grow as a nation to a very large degree depends upon our scientific progress. Moreover, it is not enough simply to keep abreast of the rest of the world in scientific matters. We must maintain our leadership.[1]

President Harry Truman spoke those words in 1950, in the aftermath of World War II and in the midst of the Cold War. Indeed, the scientific and engineering leadership of the United States and its allies in the twentieth century played key roles in the successful outcomes of both World War II and the Cold War, sparing the world the twin horrors of fascism and totalitarian communism, and fueling the economic prosperity that followed. Today, as the United States and its allies once again find themselves at war, President Truman's words ring as true as they did a half-century ago. The goal set out in the Truman Administration of maintaining leadership in science has remained the policy of the U.S. Government to this day: Dr. John Marburger, the Director of the Office of Science and Technology (OSTP) in the Executive Office of the President made remarks to that effect during his confirmation hearings in October 2001.[2]

The United States needs metrics for measuring its success in meeting this goal of maintaining leadership in science and technology. That is one of the reasons that the National Science Foundation (NSF) and many other agencies of the U.S. Government have supported the World Technology Evaluation Center (WTEC) and its predecessor programs for the past 20 years. While other programs have attempted to measure the international competitiveness of U.S. research by comparing funding amounts, publication statistics, or patent activity, WTEC has been the most significant public domain effort in the U.S. Government to use peer review to evaluate the status of U.S. efforts in comparison to those abroad. Since 1983, WTEC has conducted over 60 such assessments in a wide variety of fields, from advanced computing, to nanoscience and technology, to biotechnology.

[1] Remarks by the President on May 10, 1950, on the occasion of the signing of the law that created the National Science Foundation. *Public Papers of the Presidents* 120: p. 338.

[2] http://www.ostp.gov/html/01_1012.html.

The results have been extremely useful to NSF and other agencies in evaluating ongoing research programs, and in setting objectives for the future. WTEC studies also have been important in establishing new lines of communication and identifying opportunities for cooperation between U.S. researchers and their colleagues abroad, thus helping to accelerate the progress of science and technology generally within the international community. WTEC is an excellent example of cooperation and coordination among the many agencies of the U.S. Government that are involved in funding research and development: almost every WTEC study has been supported by a coalition of agencies with interests related to the particular subject at hand.

As President Truman said over 50 years ago, our very survival depends upon continued leadership in science and technology. WTEC plays a key role in determining whether the United States is meeting that challenge, and in promoting that leadership.

Michael Reischman
Deputy Assistant Director for Engineering
National Science Foundation

Table of Contents

List of Figures

List of Tables

Preface

This report was prepared by the World Technology Evalutation Center (WTEC), a nonprofit research institute funded by grants and other awards from most of the Federal research agencies. Among other studies, WTEC has provided peer reviews by panels of U.S. experts of international research and development (R&D) in more than 55 fields since 1989. In 2004, WTEC was asked by several agencies to assess international R&D in sytems biology. This report is the final product of that study.

We would like to thank our distinguished panel of experts, who are the authors of this report, for all of their efforts to bring this study to a successful conclusion. We also are very grateful to our foreign hosts for their generous hospitality, and to the participants in our preliminary workshop on U.S. Systems Biology R&D. Of course, this study would not have been possible without encouragement from our sponsor representatives: Bruce Hamilton, Frederick Heineken, Carol Lucas, and Maryanna Henkart (NSF); Sri Kumar (DARPA); Marvin Frazier (DOE); Richard Wiggins (EPA); Stephen Davison (NASA); Dan Gallahan and Peter Lyster (NIH); and Angela Hight-Walker (NIST).

This report covers a broad spectrum of material on the subject, so it may be useful to give a preview here. The Executive Summary was prepared by the chair, Marvin Cassman, with input from all the panelists. The chapters in the body of this report present the panel's findings in an analytical organization by subdiscipline. Appendix A provides the biographies of the panelists. Appendices B and C contain the panel's individual reports on each site visited in Europe and Japan, which form a chronological or geographic organization of much of the material. To establish a baseline for comparison, a workshop to report on U.S. research was held. A summary of the workshop is at Appendix D. Finally, a glossary is provided as Appendix E.

All the products of this project are available at http://www.wtec.org. The full-color electronic version of this report is particularly useful for viewing some of the figures.

Michael J. DeHaemer
Executive Vice President
WTEC, Inc.

CHAPTER 1

Executive Summary and Introduction

Marvin Cassman

BACKGROUND

Systems biology has become a major force in the past five to seven years. As with all new developments in science, the emergence of new approaches is a result of limitations in the existing model, in this case the limitations of molecular biology. For the past 40 years the paradigm for predicting phenotype has focused on single gene defects. This extraordinarily powerful approach has been the major contributor to an understanding of the function of individual genes and proteins. It seems less likely that it will yield an understanding of complex biological behavior, from individual cellular activities such as motility to the operation and integration of organ systems.

The current textbook image of biological processes is that of a static model of loosely linked, highly detailed, molecular devices. However, every biologist knows that dynamic processes drive biology. The physiologist Walter Cannon provided a clear statement of this concept in 1932 when he coined the term "homeostasis." One contemporary definition of a homeostatic system is "… an open system that maintains its structure and functions by means of a multiplicity of dynamic equilibriums rigorously controlled by interdependent regulation mechanisms." This description neatly encapsulates several of the key issues in systems biology—dynamic processes, interdependent regulatory controls, and the operation of multiple interacting components. The inability of most of modern biology to take these behaviors into account is certainly a major reason why the function and malfunction of complex biological processes is still poorly understood, despite biologists' increasingly detailed knowledge of the

M. Cassman et al. (eds.), Systems Biology, 1–13.

components of these processes. Systems biology brings the promise, if not yet the reality, of offering a more complete understanding of health and disease. (It is striking that at about the same time as Cannon was developing the concept of homeostasis, H. S. Black was refining the concept of negative feedback control at Bell Labs. The current active involvement of engineers in systems biology can be considered to be a convergence of these two threads of thought.)

All this interest has emerged despite the inability to arrive at a consensus definition of systems biology. Elements that appear in virtually all definitions are "networks," "computation," "modeling" and often, "dynamic properties." For the purposes of this study the objective of systems biology has been defined as the understanding of network behavior, and in particular their dynamic aspects, which requires the utilization of mathematical modeling tightly linked to experiment. This involves a variety of approaches, such as the identification and validation of networks, the creation of appropriate datasets, the development of tools for data acquisition and software development, and the use of modeling and simulation software in close linkage with experiment, often done to understand dynamic processes. Of course, the definition becomes ambiguous at the margins. But at the core is the focus on networks, which makes it clear that the goal is to understand the operation of the systems, rather than the component parts. It also tries to distinguish between systems biology and what the WTEC study panel defined as systematic biology. Systematic biology can be considered the large-scale, high-throughput collection of specific data sets and their organization and interpretation, usually through the application of advanced bioinformatics tools. Clearly, this is at one boundary of systems biology, since the data sets may be used for systems analysis. Although these two very recent approaches to understanding biology are closely linked and have developed over more or less the same time frame, they are quite distinct. Systematic biology is a consequence of the enormous success of the genome program, together with the development of new technologies for the high-throughput collection and analysis of data on individual molecular events. In this respect it is a lineal descendent of the primary driver for biology in the second half of the 20th century, molecular biology/genetics. In contrast, systems biology has a much older lineage, and can be considered a variant of computational physiology, one example of which is the Hodgkins-Huxley equation of 1952. More recent approaches include the development of metabolic control analysis in the 1970s (Kacser, Burns, 1973; Heinrich, Rapoport, Rapoport, 1977) and detailed anatomic- and molecularly based models of cardiac function (Kohl, Noble, Winslow, Hunter, 2000). These approaches, as well as systems biology, are distinguished from earlier attempts at systems modeling in that they incorporate the detailed molecular knowledge that has been and continues to be generated. However, they all assume that an understanding of physiological functions (phenotype) requires knowledge of the behavior of systems or networks.

At the heart of systems biology is the need to couple advanced modeling and simulation with experiment. This thrusts biology into a new era. For the past century the tools and concepts of chemistry have driven biology. This is so much a part of the landscape of modern biology that it will be shocking to many when, in the next century, biology becomes largely driven by engineering and physics. This is a consequence of the fact that understanding the dynamics of even the simplest biological networks requires the application of mathematical approaches and the generation of models and simulations. These mathematical tools are not now part of the average biologists' training. Indeed, biology has almost become the province of those who want to do science without learning mathematics. Consequently, two major issues in the evolving field of systems biology are needed to create functional collaborations between engineers, physicists, and biologists, and to produce a new generation of scientists that will be conversant with both the mathematical tools and the biological systems. Both of these issues were a major concern of the study.

Having said all this, it must be stressed that systems biology is still a very new field. Although individual investigators have for some years been studying the properties of biological networks using quantitative approaches, until recently they were few, relatively isolated and, to a significant extent, ignored. In part, this was due to a long-standing bias against modeling and simulation, particularly in cell biology. This bias was not completely irrational, since many modeling approaches had very little connection to experiment and consequently rarely told biologists anything they wanted to know. In contrast, the close ties of computation and modeling with experiment distinguish modern systems biology. This will be repeated often in the following chapters. It is also worth repeating Michael Faraday's comment, "All this is a dream. Still, examine it by a few experiments. Nothing is too wonderful to be true, if it be consistent with the laws of nature, and in such things as these, experiment is the best test of such consistency" (Hamilton, 2005).

The investment by government agencies of millions of dollars of federal funding since 1998 demonstrates the increasing interest in systems biology. In addition, the major journals have devoted special issues to the subject. A new journal of systems biology, called IEE Systems Biology, and most recently an online journal entitled "Molecular Systems Biology" published jointly by EMBO and Nature Publishing Group, have recently been established. Institutes have sprung up, numerous meetings are held, and increased funding can be observed both in the U.S. and abroad. Yet core issues affecting the progress of the field remain to be resolved, and some of these will be addressed in this volume. If systems biology is not quite yet a discipline, it is clearly more than a fad.

A final caveat is that limitations of time and resources prevented us from visiting many important research sites around the world. For example, we regret that we were not able to see the work underway in the Scandinavian countries and Israel, as well as Asian nations such as Singapore and Korea. Additionally, we missed numerous laboratories even in the countries we visited, and consequently some important research areas were surely neglected.

OBJECTIVES OF THE WTEC STUDY

The recent growth in interest in systems biology has not been accompanied by a systematic evaluation of activities in the U.S. and abroad. Led by the National Science Foundation, a number of U.S. governmental agencies involved in the support of research asked the World Technology Evaluation Center (WTEC) to conduct a study of systems biology activities in the U.S. and abroad to support important policy and funding initiatives. The goals are:

- To understand the state of current research
- To determine what is needed to support future research
- To understand the opportunities for international collaboration

The United States needs knowledge of and access to the latest international developments in this field in order to proceed expeditiously with promising applications in this rapidly developing field. The number and diversity of its sponsors reflect the breadth of interest in this study. These include the National Science Foundation, the Department of Energy (DOE); the Defense Advanced Research Projects Agency (DARPA) of the Department of Defense; the National Aeronautics and Space Administration (NASA); the National Cancer Institute (NCI) and the National Institute of Biomedical Imaging and Bioengineering (NIBIB) of the National Institutes of Health; the National Institute of Standards and Technology (NIST); and the Environmental Protection Agency (EPA).

PANEL MEMBERS

- Marvin Cassman, San Francisco, CA (Chair)
- Adam Arkin, University of California, Berkeley
- Frank Doyle, University of California, Santa Barbara
- Fumiaki Katagiri, University of Minnesota
- Douglas Lauffenburger, MIT

- Cynthia Stokes, Entelos Corp.

STUDY SCOPE

Broadly, the scope of the study included:

- Organization and regulation of biological networks
- Tools for analyzing the spatial and temporal behavior of networks
- Approaches to the education of graduates and undergraduates in systems and computational biology
- Trends in government interest and support of systems biology programs

The report follows an outline that was first presented at a meeting with sponsors on February 23, 2004, further defined at a workshop with prominent U.S. researchers on June 4, 2004, and refined over the months of discussion and visits carried out by the panel. It reflects not only activities observed during the site visits but also research directions which, in the opinion of the panelists, were often underrepresented but require more emphasis to ensure the progress of the field. The format for this volume is:

1. Introduction and Executive Summary (Marvin Cassman)
2. Data Generation and Analysis (Fumiaki Katagiri and Adam Arkin)
3. Systems Inference (Frank Doyle and Douglas Lauffenburger)
4. Network Organization and Modeling (Cynthia Stokes and Adam Arkin)
5. Education, Infrastructure, and National Programs (Marvin Cassman, Frank Doyle, Douglas Lauffenburger)
6. Plant Science (Fumiaki Katagiri)

PLAN OF THE STUDY

The first formal discussion of the study occurred at a workshop at NSF on February 23, 2004. In addition to most of the panelists and WTEC staff, there were representatives from all of the organizations sponsoring the study. On top of a broad discussion of the goals of the study, an outline was generated for a workshop to provide baseline information on U.S. activities in systems biology. This workshop was held on June 4, 2004, at NSF.

Site visits were made to Europe and Great Britain on July 5–9, 2004, and to Japan on December 13–17, 2004. Team members visited 16 sites in the EU and Switzerland and 12 in Japan. The site reports are appended to this report.

At all the sites that were visited (see Table 1.1), the hosts treated us with the utmost consideration. The study sponsors (in particular Fred Heineken and Semahat Demir of NSF who traveled with the panel) and participants thank them for their hospitality with the hope that this volume and anything that emerges from it will prove of value to them.

Table 1.1
Sites visited in Europe and Japan

Europe		
Site	**Panelists**	**Date**
Cambridge University, Department of Anatomy	Katagiri, Lauffenburger, Stokes	9 July 2004
European Bioinformatics Institute	Katagiri, Lauffenburger, Stokes	9 July 2004
European Commission Office	Ali, Arkin, Cassman, Doyle, Heineken	8 July 2004
German Cancer Research Center (DKFZ) Heidelberg	Arkin, Cassman, Doyle, Heineken	7 July 2004
Humboldt University	Arkin, Cassman, Doyle, Heineken	6 July 2004
Max Planck Institute for Molecular Genetics	Cassman, Doyle, Heineken, Katagiri	5 July 2004
Max Planck Institute for Molecular Plant Physiology	Arkin, Cassman, Heineken, Katagiri	5 July 2004
Oxford Brookes University, School of Biological and Molecular Sciences	Lauffenburger, Stokes	6 July 2004
Oxford University, Centre for Mathematical Biology/Mathematical Institute	Ali, Lauffenburger, Stokes	5 July 2004
Oxford University, Department of Physiology	Ali, Lauffenburger, Stokes	6 July 2004
Sheffield University, Computational Biology Research Group	Katagiri, Lauffenburger, Stokes	8 July 2004
Systems*X*	Cassman	29–30 June 2004
University College London	Ali, Lauffenburger, Stokes	5 July 2004
Université Libre (Free University) De Bruxelles	Ali, Arkin, Cassman, Doyle, Heineken	8 July 2004
University of Warwick, Mathematics Institute	Katagiri, Lauffenburger, Stokes	7 July 2004
Vrije Universiteit (Free University) Amsterdam	Ali, Arkin, Cassman, Doyle, Heineken	9 July 2004

Japan		
Computational Biology Research Center (CBRC)	Arkin, Cassman, Demir, Doyle, Horning, Katagiri, Stokes	14 Dec 2004
Japan Biological Information Research Center (JBIRC)	Arkin, Cassman, Demir, Doyle, Horning, Katagiri, Stokes	14 Dec 2004
Kazusa DNA Research Institute	Cassman, Horning, Katagiri	15 Dec 2004
Keio University, Institute for Advanced Biosciences (IAB)	Arkin, Cassman, Horning, Katagiri	17 Dec 2004
Keio University, Symbiotic Systems Project	Arkin, Cassman, Demir, Doyle, Horning, Katagiri, Stokes	13 Dec 2004
Kyoto University, Bioinformatics Center	Demir, Doyle, Stokes	17 Dec 2004
Kyoto University, Cell/Biodynamics Simulation Project	Demir, Doyle, Hane, Stokes	17 Dec 2004
RIKEN Yokohama Institute	Arkin, Cassman, Demir, Doyle, Hane, Katagiri, Stokes	16 Dec 2004
Tokyo Medical and Dental University	Arkin, Cassman, Demir, Doyle, Horning, Katagiri, Stokes	15 Dec 2004
University of Tokyo, Department of Computational Biology	Arkin, Demir, Doyle, Stokes	15 Dec 2004
University of Tokyo, Institute of Medical Science	Arkin, Cassman, Demir, Doyle, Horning, Katagiri, Stokes	13 Dec 2004
University of Tokyo, Laboratory of Systems Biology and Medicine (LSBM)	Arkin, Cassman, Demir, Doyle, Horning, Katagiri, Stokes	13 Dec 2004

PRINCIPAL FINDINGS

General Conclusions

Over the past decade numerous individual laboratories around the world have been engaged in systems biology. Some of these investigators include Lauffenburger (Lauffenburger, Forsten, Wiley HS (1995), Arkin (Arkin, Ross, McAdams (1998), McAdams (McAdams, Shapiro, 1995), Leibler (Barkai, Leibler, 1997), and Savageau (Hlavecek and Savageau, 1997) in the U.S.; Bray (Levin, Morton-Firth, Abouhamad, Bourret, Bray, 1998) and Noble (McCulloch, Bassingthwaighte, Hunter, Noble, 1998) in the U.K.; Heinrich (Wolf, Heinrich, 1997), Westerhoff (Westerhof, 1995), and Goldbeter (Goldbeter, 2002) in Europe; and Kitano (Kitano, 2002), Tomita (Tomita, Hashimoto, Takahashi, Shimizu, Matsuzaki, Miyoshi, Saito, Tanida, Yugi, Venter, Hutchison, 1999), and Kanehisa (Kanehisa, 2000) in Japan. However, few of these efforts were matched by either significant national funding or institutional interest. This changed as the profusion of data and the complexity of regulatory processes mounted. The interest generated by the need to integrate molecular data into a systems approach in turn stimulated events over the last five to seven years in the U.S., and more recently elsewhere, when large investments in systems biology began to be made by national entities and research institutions.

Largely because of its head start, the WTEC panel rates the U.S. as currently ahead of the rest of the world in systems biology. The lead is reflected in the larger number of active groups, greater number of educational programs underway, and the more diverse and growing funding base. However, there is evidence of rapid development outside the U.S., much of it begun in the last two to three years. It must be stressed that the attempt to incorporate the details of molecular events obtained over the past half-century into a dynamic picture of network behavior in biological systems is only just beginning, in the U.S. and elsewhere. In particular, progress in the core activity of systems biology—modeling tied to experiment—is still limited. Successes, however defined, remain few and controversies abound. Training, research, and infrastructure all would benefit from strong international collaborations that could provide examples of novel approaches. For example, Japan and Germany have developed large-scale organizations that can address specific research issues, e.g. the Max Planck Institutes in Germany and RIKEN in Japan. The U.S. has a limited capability to create such structures and needs to develop inter-agency collaborations that will identify and support activities of this kind. Overall, the picture is of an active field in the early stages of explosive growth.

Databases and Data

The production of data and the construction of databases are visible at a roughly similar scale in the U.K., EU, and Japan, although the development of large databases in Japan was particularly striking. However, the databases examined, in the U.S. and abroad, were not always valuable to investigators developing and testing models of biological processes.

Two opposite trends exist in database organizations: large inclusive databases and small specialty databases. Both approaches have advantages and disadvantages. Large-scale databases, which primarily collect information that is not closely tied to the state of a cell, have become quite common, and their value is well understood. (How useful these data are for systems biology is less clear. In general, the degree of quantitation is too limited to be used by investigators developing and testing models of biological processes.) Standardization, although not complete, is progressing. This is not the case for many other kinds of data, particularly those tied to biological processes that are strongly conditioned by the state of the cell. It is not even clear how much "meta-data" is needed. Gene expression, protein expression, molecular localization, interactions, and post-translational modification are highly conditional. Indeed, the strain of cell used, the media, and other measurement conditions can appreciably affect the measured outcomes. There are a number of related issues, such as the amount of raw data needed, and the availability of statistical analysis and software packages used. These issues are not unique to data used for systems biology, but their absence is even more critical than in the analysis of state-independent data.

Models for data production and data storage in systems biology are highly variable, ranging from large centers with massive accumulations of high-throughput data, to small, manually curated databases. Whichever model is used, the absence of data standards that permit groups other than the producer to use, analyze, and evaluate the results is clearly a significant barrier to progress. This is an international issue, and must be solved by broad collaborative interactions.

System Inference

One finds numerous network inference studies in all of the regions described with the U.S. and Japanese (and Israeli) groups leading in the development of methodologies. All regions showed exciting application studies, with significant potential for "success stories" to emerge in the coming years.

The encouraging trends that were observed included: (i) multiple, complementary approaches to the regression of models for network inference, (ii) the incorporation of motifs and modules into network inference methods, (iii) the emergence of a nice interplay between the classical static network databases and the formats for dynamic systems biology models (*e.g.*, SBML), and (iv) the initiation of a considerable amount of curricular development in this area (notably in bioinformatics).

Of concern was the fact that the issues of: (i) explicit incorporation of dynamics, (ii) identifiability and (in)validation of models, and (iii) model iterations with design of experiment, were receiving only modest attention in the regions, with noteworthy efforts in the U.S. and Europe (particularly Germany). There were many reported examples of researchers identifying large numbers of parameters from relatively small data sets. However, there appear to be a number of groups working towards solutions to these challenges, and considerable progress can be expected in the next two to three years.

Modeling and Network Organization

Modeling and network organization analysis efforts are utilized in many areas of biological study and in all countries visited, but are definitely not ubiquitous throughout biological and biomedical research. The panel found that research efforts that closely integrated modeling with experimental work were the most productive in terms of driving new understanding of a biological system. Related to this, substantially more effort using model-based experimental design is needed to attain the data that most efficiently leads to maximally useful models. In addition, better tools for model-experiment comparison would be helpful. Significant resources are being invested in the development of modeling and simulation software worldwide, and at least some duplication of effort is apparent. Sharing of models between researchers remains a challenge but is being addressed by the development of several markup languages. Finally, the involvement and interest of industry in use of modeling in biology is significant although, again, not ubiquitous.

Plant Systems Biology

Progress of systems biology research in the plant field has been slow. However, some advanced studies shed light on unique aspects of plants. Several actions are needed to promote systems biology of plants.

To make the most out of limited funding:

- Focus on model plant species. It is clear that the majority of advanced studies have been performed with model plant species, such as *Arabidopsis*.
- Cooperate rather than compete at the global level.

To compensate for the bias against promotion of systems biology in the research community:

- Implement a sustaining, targeted funding program in plant systems biology.

To raise the next generation of researchers:

- Train biology-major students in quantitative science.
- Recruit students oriented to mathematics, engineering, physics, and chemistry into plant biology.

Education, National Programs, and Infrastructure

The future of systems biology will depend on three critical elements: education of a new generation of scientists who have both biological and mathematical training; the availability of funding that operates outside of disciplinary boundaries; and the availability of a supportive infrastructure that can accommodate the needs of an intrinsically interdisciplinary research area.

Education

The general impression is that most of the formal teaching programs, in the U.S. and abroad, are in bioinformatics rather than systems biology. Relatively few examples exist of training in modeling focused on biological systems, and where they do exist they tend to be isolated courses rather than fully integrated programs in systems biology. Most of the programs are somewhat *ad hoc* "menu selection" curricula. The difficulty of training quantitative students in biology and *vice versa* is clearly well understood and no real solution has yet been provided, although a number of experiments are underway. It is much too early to tell which, if any, of these are successful in producing qualified researchers in systems biology. Given the importance of this issue and its embryonic state, some mechanisms for exchanging information internationally and locally on best practices is essential.

National Programs

The U.S. remains one of the few countries that offers a significant targeted investment in systems biology. A clear exception is Germany, which has developed a new initiative in the systems biology of hepatocytes, beginning in January 2004. In the last few years, national programs have also been initiated in Switzerland and the U.K., and international programs at the EU level. Additionally, activities in systems biology are underway in many locations, as part of ongoing "traditional" governmental support programs. This is particularly noticeable in Japan. However, it is hard to avoid the conclusion that both the breadth and the scale of systems biology support from governmental entities are significantly greater in the U.S. than elsewhere in the world.

A possible caveat to this conclusion depends on the definition of systems biology. As noted earlier, there is a distinction between "systems biology" and "systematic biology." Systematic biology, the high-throughput collection of targeted data sets, is a booming business everywhere, fuelled by the success of the genome project. Systems biology, the computational analysis of biological networks, is much more sparsely represented. Although this is also true in the U.S., encouragement of these activities through federal funding programs is significant and growing. It was slightly discouraging to see how frequently systems and systematic biology were conflated. Although data collection is clearly critical, it was not often the case that there was a connection between the data collected and its potential use in modeling and simulation of biological systems. In general, the future of systems biology worldwide depends on the support of programs that consider experimental and data-driven approaches together with the computational methods needed to model specific biological problems. Relatively few funding programs focus explicitly on this.

Infrastructure

The infrastructure to be discussed in this study is limited to large-scale resources, specifically databases, software repositories, and centers. In order to ensure both standardization and access, it is strongly recommended that centralized resources be developed for both software and data. The third issue is the value of centers for systems biology. The creation of specialized centers is much more common in Europe and Japan than in the U.S., although the development of high-throughput centers for DNA sequencing and structural biology have proven their value. It is suggested that centers targeted to specific research problems, and specific experimental systems, could benefit systems biology in the U.S. The need for consistent and reproducible data and the need for close collaboration between theorists and experimentalists are both arguments for co-located groups that can interact easily and often. It is also far easier to enforce standards at such centers. At this point in time, systems biology can benefit from stronger centralized approaches that will allow the testing of model systems in an optimum environment.

REFERENCES

Arkin, A., J. Ross, H. H. McAdams. 1998. Stochastic kinetic analysis of developmental pathway bifurcation in phage lambda-infected *Escherichia coli* cells. *Genetics* 149: 1633–48.

Barkai, N., S. Leibler. 1997. Robustness in simple biochemical networks. *Nature* 387: 913–7.

Goldbeter, A. 2002. Computational approaches to cellular rhythms. *Nature* 420: 238–45.

Hamilton, J. 2005. "A Life of Discovery," cited in NYT Book Reviews, p. 16, March 13, 2005.

Heinrich, R., S. M. Rapoport, and T. A. Rapoport. 1977. Metabolic regulation and Mathematical Models. *Prog Biophys Mol Biol* 32: 1–82.

Hlavacek W. S., M. A. Savageau. 1996. Rules for coupled expression of regulator and effector genes in inducible circuits. *J Mol Biol* 255: 121–39.

Kacser, H. and J. A. Burns. 1973. The Control of Flux. *Symp Soc Exp Biol* 27: 65–104.

Kanehisa, M. 2000. Pathway databases and higher order function. *Adv Protein Chem* 54: 381–408.

Kitano, H. 2002. Looking beyond the details: a rise in system-oriented approaches in genetics and molecular biology. *Curr Genet* 41: 1–10.

Levin, M. D., C. J. Morton-Firth, W. N. Abouhamad, R. B. Bourret, D. Bray. 1998. Origins of individual swimming behavior in bacteria. *Biophys J* 74: 175–81.

Kohl, P, D., R. L. Noble, R. L. Winslow, and P. Hunter. 2000. Computational modeling of biological systems: tools and visions. *Phil Trans R.Soc Lond A* 358: 576–610.

Lauffenburger, D. A., K. E. Forsten, B. Will, H. S. Wiley. 1995. Molecular/cell engineering approach to autocrine ligand control of cell function. *Ann Biomed Eng* 23: 208–15.

McAdams, H. H., L. Shapiro. 1995. Circuit simulation of genetic networks. *Science* 269: 650–6.

McCulloch, A., J. Bassingthwaighte, P. Hunter, D. Noble. 1998. Computational biology of the heart: from structure to function. *Prog Biophys Mol Biol* 69: 153–5.

Tomita, M., K. Hashimoto, K. Takahashi, T. S. Shimizu, Y. Matsuzaki, F. Miyoshi, K. Saito, S. Tanida, K. Yugi, J. C. Venter, C. A. Hutchison 3rd. 1999. E-CELL: software environment for whole-cell simulation. *Bioinformatics* 15: 72–84.

Westerhoff, H. V. 1995. Subtlety in control—metabolic pathway engineering. *Trends Biotechnol* 13: 242–4.

Wolf, J., R. Heinrich. 1997. Dynamics of two-component biochemical systems in interacting cells; synchronization and desynchronization of oscillations and multiple steady states. *Biosystems* 43: 1–24.

CHAPTER 2

Data and Databases

Fumiaki Katagiri and Adam Arkin

INTRODUCTION

Data does not appear to be in short supply in contemporary biology. The development of high-throughput technologies, in particular, has generated massive amounts of information. While these technologies produce information about the chemistry of a system, as with the sequence and structure databases, the biological status of the organism is often of little importance. However, when the goal is describing a process, such as signal transduction or gene expression, the information gathered will be highly conditioned by cell type, experimental conditions, and other variables. Since the modeling of dynamic biological processes is a central aspect of systems biology, the nature of the data available to create and test models is of great importance. This chapter will examine some aspects of data generation and data storage and access, as it applies to systems biology. It will not focus on the technologies themselves. Systems biology relies on a wide variety of data types. Some of these have become fairly standard, such as sequence and structure information, mass spectrometry, and microarrays. Others are still in the process of development, such as single cell measurements using multicolored fluorescent assays, fluorescent antibodies and covalent conjugates, and quantum dots using high-resolution microscopy and flow cytometry. The diversity of tools used precludes a detailed discussion in this chapter. Additionally, although broad access of all tools to investigators is an issue, there do not appear to be systematic differences between countries in the availability of technologies.

M. Cassman et al. (eds.), Systems Biology, 15–30.

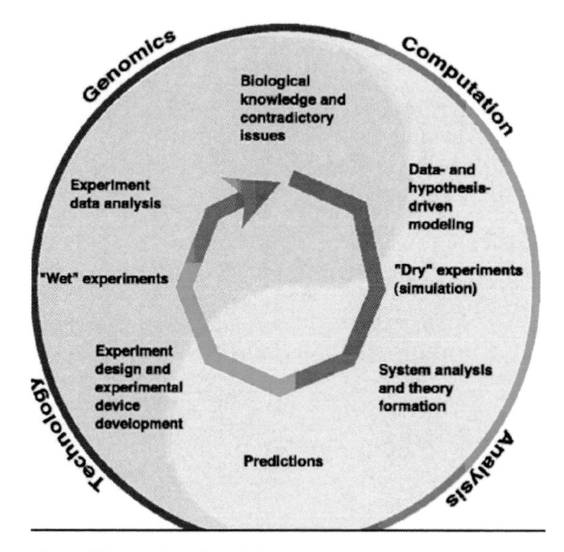

Figure 2.1. Hypothesis-driven research in systems biology (Kitano, 2002).

The interaction of experiment and models requires accurate datasets to infer network structures, to create the models, and to test and distinguish the predictions from multiple models. Successive iterations of model building, prediction, experiment, and subsequent refinements of the models are the result. This repeating cycle of experimentation and theoretical work is the engine that pushes the progress of systems biology research (Kitano, 2002). Consequently, while the concept of systems biology is not all that new, one reason that it has displayed renewed vigor is the impressive advance of experimental technologies. The development and spread of various approaches to high-throughput measurements have been contributing to generation of a large amount of systematic data from a wide variety of biological systems. These data have fueled high-throughput discovery research based on reductionist approaches. Furthermore, rapid generation of such data has given people a sense of hope that we may now be able to collect sufficient information to understand biological phenomena as behaviors of dynamic systems. Is this hope built on a solid foundation?

Roughly speaking, we can distinguish two stages in systems biology research. The type of experimentations useful in each stage is also distinct. When the network structure of interest is not well-defined, systematic and broad-spectrum characterizations, such as global profiling, provide the necessary information. When the network structure is well defined and quantitative models are built based on the network structure, the experimental information demanded is very specific to the network of interest, the proposed models, and the questions asked about the network. In such a case, the data collection is model-driven and experiments need to be designed according to the specific demands. In addition, the techniques used must, in most cases, provide quantitative results.

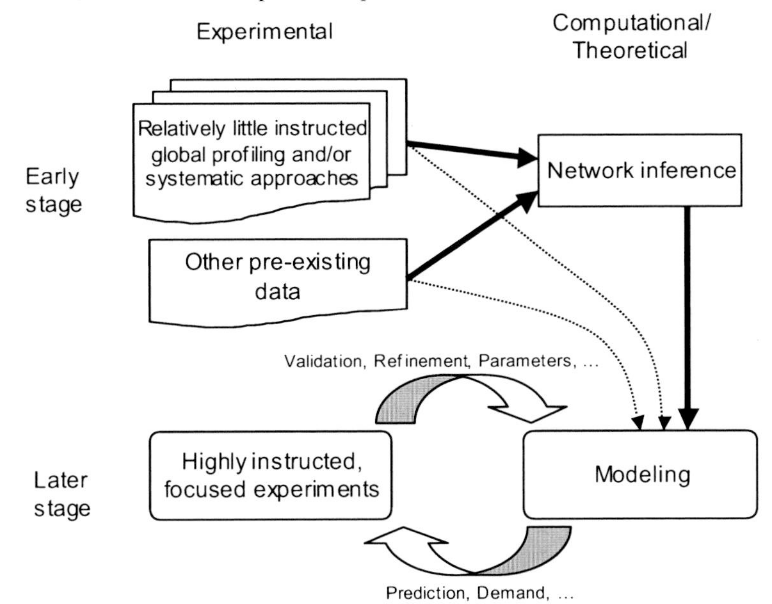

Figure 2.2. The two different stages in systems biology study and the types of useful experimental approaches for each stage.

In both types of experimentation, it is crucial to use a combination of methods to perturb the network and to measure the effects of the perturbations. When no model is available, there is little *a priori* information about the network to limit the set of targets for measurement. Consequently, the ideal is to densely cover the space of all the possibilities in both perturbation and measurement. The ideal perturbation method thus enables perturbation of every node or edge in the network specifically and quantitatively, and the ideal measurement methods measure all the parameters that define the state of the network. Practically, systematic genetic perturbation is often used as a perturbation method. Ideally, for the initial phase of determining overall regulatory organization and inferring what should be in the model, four criteria are important. First, the data should be exhaustive within each category of measurements, such as the measurement of messenger ribonucleic acid (mRNA) levels, measurement of protein levels, measurement of cell sizes, etc. For example, when mRNA levels are measured, the mRNA levels of all the genes in the organism should be measured. If the measurement must be limited to a subset of the genes, we should have a good idea that the subset likely contains all the genes that are important for the network of interest. Second, many different categories, ideally all the possible categories, of data should be collected in a correlated manner. Correlative measurements in different categories are crucial for integration of various measured events into a single network because such data are context-dependent. For example, if a protein-protein interaction measurement is performed under one condition, the measurement may not be useful under a different condition. Third, the data should have sufficient resolution in time and space. If a measurement does not have sufficiently dense time points, information about the dynamics of the system is limited. Considering an example of spatial resolution, if a measurement is performed with a mixture of different cell types that behave differently, the resulting averaged measurement fails to detect distinct cell-type modules. Fourth, the measurement should be quantitative. Whereas binary Boolean models have their own utility in many cases, they are often not sufficient to capture important network dynamics.

One major trend that is initiated by genomics research is the development of various highly parallel measurement (broad profiling) technologies. RNA profiling methods allow reasonably complete mea-surements in many organisms in which the genome sequences are known. The use of microarrays and reverse transcription-polymerase chain reaction (RT-PCR) results in good sensitivity and accuracy. Protein and metabolite profiling methods are improving quickly by combining chromatographic or electrophoretic separation methods with mass spectrometry-based methods. Although our study did not cover them, microfluidics and other micromanufacturing technologies are expected to dramatically improve the cost, speed, and labor-intensiveness of highly parallel measurements. However, we are still missing good profiling methods for many categories of data. For example, if we want to know the amount of a particular modified form of a particular protein in a particular subcellular location, we still need to perform focused research. And these data types are often the most useful for mechanistic modeling of pathways. Another issue with broad profiling technologies is that they do not provide sufficient accuracy in many cases. In addition to technical challenges, practical issues, such as limitations in budget, time, and human resources, could limit the kinds of data obtainable by broad profiling methods.

The discussion in this chapter will be on both large-scale data collection and profiling technologies as well as smaller, model-focused data. However, the more global databases, which are largely independent of the biological state of the systems, clearly predominate. It must be noted that there are remarkably few examples of the melding of modeling and measurement technology in the U.S. or Japan, and the large-scale databases are rarely useful for systems biology studies. In Europe, this melding is an essential feature of EU-funded research consortium projects (COMBIO, COSBICS, DIAMONDS, EMI-CD). A new initiative has been started

(ENFIN) to apply the large databases to systems biology research. Incompleteness of information, e.g., limitations in quantitation, accuracy, resolution, and the categories of data, is the major reason that data generated by broad profiling technologies are underutilized in systems biology research. A large number of data points with limited measurement accuracy also present challenges in having sufficient statistical power in analysis. Applications of detailed systems biology modeling are so far often limited to relatively small, well-isolated networks in which it is practical to perform targeted, intensive experimentation to obtain key information to answer specific questions.

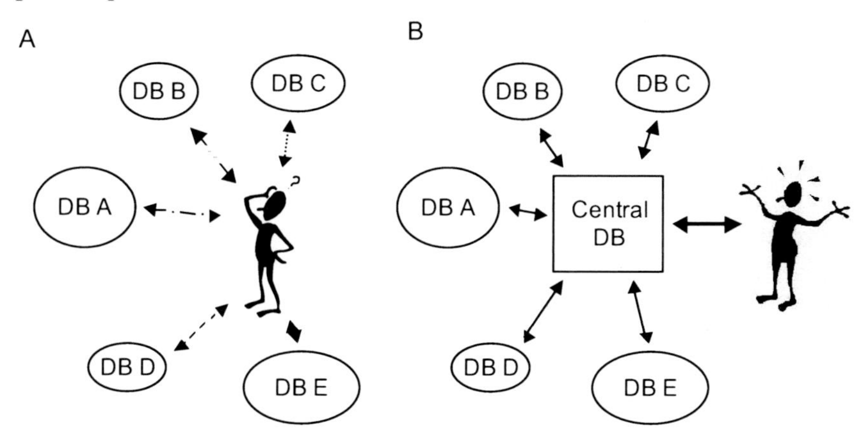

Figure 2.3. The importance of standardization and virtual consolidation of databases.

Exhaustive, quantitative measurements with high resolution in many categories necessarily generate a large amount of data. To effectively integrate such data into models of biological systems, powerful database platforms are essential. Archiving quickly growing, context-dependent data in a usable manner is a serious challenge. Curating a large amount of diverse information for many different systems requires a wide range of knowledge. These situations tend to drive the generation of many, relatively small, highly focused databases. However, having many small databases presents a funding challenge for maintaining and updating such databases as well as creating a standardization issue. Standardization is crucial in not only maintaining the integrity of the entire body of related databases but also in making them usable to general users. It is also an issue in the peer review of results and methods.

We will compare activities in producing and archiving data of importance to systems biology in the U.S., Europe, and Japan. This includes not only scientific aspects but also social background and funding situations that may have had significant influence in shaping the research activities in these regions. We will also consider general database issues apart from regional considerations, since the main issue is standardization at the global level. In the end, we will point out specific needs and recommend some possible directions for the experimental part of systems biology research.

RESEARCH ACTIVITIES IN THE U.S.

Databases

Large-scale technologies began with the sequencing programs directed to the understanding of the human genome. Following on their successes there emerged a variety of genome-wide measurements of cellular function. These include whole genome gene expression microarrays, large-scale deletion libraries, structure, and the application of mass spectrometry to assess, for example, protein-protein interactions. These in turn have led to the creation of databases containing a variety of data types, including organism-specific databases supported by federal funding, usually from the National Institutes of Health (NIH), such as Saccharomyces cerevisae (http://www.yeastgenome.org); WormBase (http://www.wormbase.org/) for C.elegans and other nematodes; FlyBase (http://flybase.bio.indiana.edu/) for Drosophila melanogaster; and The Arabidopsis Information Resource (http://arabidopsis.org/) for Arabidopsis thaliana. These databases include a diverse array of data types, such as fitness of deletion mutants (the yeast database) and anatomy and spatial expression data (FlyBase and WormBase). They represent significant ongoing investments in the United States in centralizing and standardizing data necessary for systems biology. The co-evolving efforts associated with this are the somewhat spotty and erratic investments in data standards such as the microarray data standards outlined at the Microarray Gene Expression Data Society (MGED) (http://www.mged.org/); ontologies for describing gene function or structure with a controlled vocabulary (http://www. geneontology.org/); and pathway information (http://www.biopax.org/) and biological model (http://sbml.org/index.psp) storage and transport.

At the next level from these comprehensive databases are national investments in smaller, more focused, but still relatively large-scale projects. These include support from the National Institute of General Medical Sciences of the NIH for so-called "Glue Grants" which require coordination of multi-site activities around specific targeted goals, in many of which systems biology plays a role. These include the Cell Motility Consortium (http://www.cellmigration.org) dedicated to developing rea-gents, measurements, and models of cell migration; the lipid MAPS Consortium (http://www.lipidmaps.org/), to identify, quantitate, and define the interactions of cellular lipids; the Consortium for Functional Glycomics (www.functionalglycomics.org/), to understand the role of carbohydrates in cell-cell communication; and the Alliance for Cellular Signaling (http://www.signaling-gateway.org/), to examine signal transduction pathways. Other such tightly focused efforts include the Alpha Project (http://www.molsci.org/), supported by the National Human Genome Research Institute, to look at the pheromone response in yeast and the Department of Energy's Genome to Life (GTL) projects that seek to understand the mechanisms by which microbes function in the environment.

Data Analysis

Much of the technology that was at one time thought to be too expensive for individual investigators (and thus needed to be centralized) has, in fact, found its way into many laboratories. The most prevalent of these are microarrayers, but there are also improved fluorescent microscopes, small flow cytometers, and even mass-spectrometers that have come down in price far enough such that a single lab can own one. In many cases, the experiments for any particular system need to be optimized such that even a university central facility would be inappropriate (unlike sequencing, which can be easily outsourced). Thus, the rate of data generation (and the diversity of experimental protocols and reporting formats) has exploded and driven a large boom in the academic (and industrial) data analysis efforts. These include the development of statistical methods and experimental designs specialized in ferreting out sources of systematic and random error in microarray experiments, typified by the statistically grounded methods of Trevor Hastie and Robert Tibshirani at Stanford (Hastie and Tibshirani, 2004), and Terry Speed at the University of California Berkeley (Bolstad et al., 2003). It extends to sophisticated methods for correlation and fusion of data across experimental conditions and different measurement types to derive the "modules" of co-regulation and their relationships, typified by the developments in statistical graph modeling approaches championed by Daphne Koller (Stanford) (Segal et al., 2003), David Gifford (MIT) (Bar-Joseph et al., 2003), and Michael Jordan (U.C. Berkeley) (Lanckriet et al., 2004). With these analytical methods as a foundation, together with data on the upstream sequence of co-expressed genes, and knowledge of protein structural domain interactions, these correlative methods are being applied to the network inference problems described in the third chapter. Additionally, John Doyle at Caltech (Carlson and Doyle, 2002), Frank Doyle at University of California, Santa Barbara (UCSB) (Gunawan et al., 2005), and Michael Frenklach (Rao et al., 2004) and Andrew Packard at U.C. Berkeley are all beginning to employ statistically and physically grounded data analyses for full dynamic model parameterization, model validation and model discrimination. These methods are expected to become more important as more quantitative data and models are developed and need to be formally compared.

Modeling and Data Collection

The Alpha Project and the Alliance for Cell Signaling represent two attempts to systematically collect data in well-defined systems together with efforts to create models to predict system behavior. There are a relatively small number of laboratories that are engaged in developing models and concomitantly testing them experimentally. The systems studied are quite diverse, ranging from bacterial chemotaxis (Alon et al., 1999) to the epidermal growth factor (EGF) receptor in fibroblasts (Wiley et al., 2003) to the Wnt signaling pathway in *Xenopus* (Lee et al., 2003). Since each of these represent a different biological system, a centralized database is not easy to justify for this kind of data collection. However, this does not mean that standardization and easy access to the data and computational methods is not required. Indeed, it is essential to effectively review manuscripts and develop statistics and algorithms to analyze and cross-compare experiments from a variety of laboratories. In whatever form, databases need to be established that conform to common standards of data collection and ontology, and that contain enough meta-data defining conditions of measurement to allow such evaluation and comparisons. This will be discussed in more detail later.

RESEARCH ACTIVITIES IN EUROPE

The panel saw a variety of styles of organizations in data and database aspects of systems biology research in Europe in July 2004. They include large institutes that are directed toward common goals, large consortiums, and small groups. The driving forces of different organizations are also often different. In some cases, the leadership of particular individuals was the key. In other cases, funding initiatives were the major factors.

Databases

Large-scale databases of sequence and structure are as common in Europe and the U.K. as in the U.S. One example is the European Bioinformatics Institute (EBI) located in Cambridge (http://www.ebi.ac.uk/). The Institute's focus used to be database technologies that handle and facilitate use of a large amount of data generated by genomics research. The focus has been shifting toward studies aimed at biological functions while taking advantage of the Institute's strength in computer technology. The Institute now includes research groups conducting biological experimentation. This shift from a purely computational institute to an institute with capabilities in both computation and experimentation reflects the recognition by computer scientists of the importance of close integration between experimental and theoretical work.

A large, focused operation is generally best conducted at a single site. Typically, this type of operation is led by a strong leader with a clear vision of the goals to be achieved. For example, the Max Planck Institute for Molecular Plant Physiology (http://www.mpimp-golm.mpg.de/) is focused on collecting correlated data, such as expression and metabolite profiles, from a large collection of genetically perturbed Arabidopsis plants. Usually, each department in a single Max Planck Institute operates independently. However, with the strong leadership of Dr. Willmitzer, this Max Planck Institute has been shaped toward a common goal of understanding plant metabolism. Two major departments led by Drs. Willmitzer and Stitt, together with other departments, cooperatively conduct these large data generation and analysis operations.

A large consortium can also be organized with geographically dispersed laboratories. The Hepatocyte Alliance in Germany was driven by the Federal Ministry of Education and Research (BMBF) funding initiative "Systems for Life-Systems Biology." The alliance has 25 total participating groups. Funding of €14 million is provided over the three years beginning in 2004. Standardization of biological materials within the alliance was rigorously implemented. The alliance works on a number of sub-projects, including detoxification/dedifferentiation and regeneration. In the regeneration sub-project, experiments in different groups are organized according to different signaling pathways. The approach of segmenting the project into smaller, defined subnetworks allows the groups working on different projects to be moderately independent while still effectively interactive despite geographic separation among the consortium members. The effectiveness of such an operation in coordinating the production and maintaining the quality of data is still unclear, since it was only a few months underway at the time of our visit. The benefits of this approach are the inclusion of a number of high-quality groups with specialized expertise that may not be available at a single site. However, dividing this type of work among multiple sites makes quality control of resources and data more difficult and increases the chances of mistakes/accidents in tracking them. The overall cost- and time-efficiency in production is also lower.

Modeling and Data Collection

As in the U.S., there are relatively few groups with closely linked efforts in modeling and experimentation where we could identify systematic data collection efforts. However, one prominent example is the group headed by Denis Noble at Oxford. They have a long history of studying the heart using experimental and modeling approaches at multiple scales from molecular to physiological aspects. While his laboratory has generated a large amount of data, specific demands of the model often call for different expertise. Therefore, various collaborations were developed at different stages of the research (http://www.physiome.org/), notably including the anatomic studies of Peter Hunter in New Zealand (http://www.bioeng.auckland.ac.nz/home/home.php). The data collections have been specifically linked to the needs of the model.

RESEARCH ACTIVITIES IN JAPAN

Funding for basic research by the Japanese government has dramatically increased in the last decade. Research in genomics, related high-throughput discovery research, and the bioinformatics supporting them has benefited from the increased financial support. Japan is particularly strong in the development of large databases.

Databases

The panel visited several large research institutes with strong database components during our visit in December 2004. They include the RIKEN Yokohama Institute (http://www.yokohama.riken.jp/indexE.html), which represents an effort under the Ministry of Education, Culture, Sports, Science, and Technology (MEXT). Among the databases supported by RIKEN is the Mouse Genome Encyclopedia (http://genome. gsc.riken.go.jp/), which collects the sequences, physical clones, gene expression profiles, and protein-interaction data generated by members of the FANTOM (Functional Annotation of Mouse complementary deox-yribonucleic acid (cDNA)) consortium. RIKEN also has the Arabidopsis Genome Encyclopedia (http://rarge.gsc.riken.go.jp/), which includes collections of mutants and full-length cDNA clones, microarrays, and shape phenotypes of the mutants. In the next phase the Arabidopsis project will integrate genome to phenome and metabolome. In 2004 RIKEN initiated the Genome Network Project (http://www.mext-life.jp/genome/english/index.html) to identify transcriptional regulatory networks in the human genome. This is a national project, where the biological projects and technology development is performed by 12 groups of independent investigators while RIKEN provides the core data and resource production capability.

Dr. Shibata's group at the Kazusa DNA Institute, which is mainly funded by Chiba prefecture, leads a collaborative effort to investigate metabolic networks in plants (Hirai et al., 2005). This project is also supported by NEDO (New Energy and Industrial Technology Development Organization), which is operated by the Ministry of Economy, Trade, and Industry (METI) and focuses on Arabidopsis. A particularly interesting database developed by Dr. Kanaya (Nara Institute of Science and Technology) is directed to candidate compound identification, and includes 20,000 microbial compounds and 100,000 plant compounds, and contains a variety of compound information as well as the species in which a particular compound has been detected.

The Japan Biological Information Research Center (JBIRC) is part of the National Institutes of Advanced Industrial Science and Tec-hnology (AIST), which operates under METI. It includes an integrated human genome annotation database (H-Invitational DB, http://www.h-invitational.jp/), containing information on 41,118 full-length cDNA clon-es including gene structures, functions, domains, expression (in some cases), diversity, and evolution.

Finally, mention must be made of the KEGG (Kyoto Encyclopedia of Genes and Genomes) database (http://www.genome.jp/kegg/) of metabolic networks developed and maintained by Dr. Kanehisa's group at Kyoto University, which is widely used throughout the world.

The infrastructure to support these efforts is uniformly outstanding and frequently astonishingly good.

MODELING AND DATA COLLECTION

Most of the data development efforts come from high-throughput discovery research and technology development, and, once again, the number of groups that focus on the interaction of models and experiments in systems biology is still small in Japan as elsewhere. However, there are a number of examples of this kind of activity. Under the strong leadership of Dr. Kodama, the Laboratory of Systems Biology and Medicine at the University of Tokyo (http://www.lsbm.org/site_e/univ/) is streamlined for discovery and production of diagnostic and therapeutic antibodies. As part of this Dr. Ihara's group developed a protein interaction database through text mining to help understand mechanisms and select protein targets against which to raise antibodies. The Symbiotic Systems Project (http://www.symbio.jst.go.jp/symbio2/index.html) headed by Dr. Kitano has a number of programs that closely tie experiment and models. He is also working with a consortium that involves developing a database of automated recording of cell lineage in C. elegans mutants. The Institute for Advanced Biosciences (IAB, http://www.iab.keio.ac.jp/) at Keio University in Tsuruoka was built to fulfill Dr. Tomita's vision of the experimental needs of systems biology, so large-scale experimentation in the Institute is closely connected with theoretical work in the study of *E. coli* metabolism. In this project, Dr. Mori's collection of systematic mutants of E. coli is an outstanding resource (Mori, 2004). Dr. Ueda at the RIKEN Center for Developmental Biology (CDB) is a young PI building a research program for study of circadian rhythms in mammalian cells that involves high-throughput measurements and theoretical work (Ueda et al., 2005).

Many of our Japanese hosts acknowledged that much of their research does not fit our definition of systems biology. However, this situation may be changing. Dr. Yao, who is a consultant for the RIKEN Yokohama Institute, JBIRC, and CBRC, reported in our final workshop the MEXT plan for support focused on systems biology research. MEXT clearly considers systems biology as a next step after the establishment of high-throughput research infrastructure, which has been heavily funded in the past decade. If the program is well implemented, especially by facilitating involvement of more modeling-type researchers, Japan, with a high level of infrastructure, has great potential to make rapid progress in systems biology research.

Conclusions

There are two opposite trends in database organizations: large inclusive databases and small specialty databases. Both approaches have advantages and disadvantages. Large-scale databases, which primarily collect information that is not closely tied to the state of a cell, have become quite common, and their value is well understood (How useful these data are for systems biology is less clear. In general, the degree of quantitation is too limited to be used by investigators developing and testing models of biological processes). Standardization, although not complete, is progressing. This is not the case for many other kinds of data, particularly those tied to biological processes that are strongly conditioned by the state of the cell. It is not even clear how much "meta-data" is needed. Gene expression, protein expression, molecular localization, interactions, and post-translational modification are highly conditional. Indeed, the strain of cell used, the media, and other measurement conditions can appreciably affect the measured outcomes. There are a number of related issues, such as the amount of raw data needed, and the availability of statistical analysis and software packages used. These issues are not unique to data used for systems biology, but their absence is even more critical than in the analysis of state-independent data.

As noted in the introduction to this chapter, the variety of biological systems used for modeling in systems biology tends to drive the creation of many small, highly focused databases. The size of a database and the level of manual curation are inter-dependent. One researcher cannot be an expert in many different biological systems, and thus intensively and manually curated databases tend to be small, with good quality control of the data. However, small databases are often independently developed and have substantial overlaps with other databases, and they are not well standardized. The absence of standardization severely limits the utility of these databases. For researchers who are not very familiar with a particular biological system, it may not be very easy to find the most appropriate database for their purposes. Furthermore, it is difficult for a small database project to be continuously funded for maintenance and updates.

The major reason that manual curation is valuable in databases is that some types of data are not easily formatted according to fixed rules. This is evident when a primary source of data is not designed for transferring the data to databases. Descriptive data from literature is a typical example. Some efforts must be made to deal with this issue. First, multiple different terms can be used to describe the same thing, or, even worse, the same terms can be used to describe different things, depending on the context. Efforts to impose a controlled vocabulary, i.e. ontology projects, were initiated to make data stored in databases self-consistent. If a controlled vocabulary is imposed at the stage of generating primary data sources (e.g. literature), the difficulties of manual curation in collecting data will be eased. Second, the relationships among terms and the context in which terms are used are very difficult to automatically capture while they are crucial in collecting accurate data from literature. One option would be to have authors of a paper submit a formatted, database-friendly summary of the work at the submission of the manuscript. Third, the data needs to be in a form that will allow critical evaluation of its reliability. This function of expert curation is becoming more important as more data are becoming available which are not carefully quality-controlled.

For convenience to users of databases, it is desirable to have small databases virtually consolidated in a large database framework using the same standards. In other words, accessing multiple databases through a central database should seem almost seamless to users. In this way, it could be possible to maintain small databases after termination of their funding although updating them could still remain a challenge. A high level of integration could be difficult with already highly developed databases due to the difference in underlying database schemes. However, if we set general standards, it could be achieved with newer, small databases. It would be helpful if such hierarchical relationships of standardized databases were organized in a research community for each particular biological system and research field.

In summary, progress in systems biology requires a balance between bottom-up efforts, which are based on creativity and competition/ collaboration of individual researchers and/or laboratories, and top-down organization efforts, which are necessary for standardization and efficient use of limited resources. Generally speaking, the current success of U.S. research is largely owed to emphasis on bottom-up efforts. However, in many aspects of experimental/data technology for systems biology, we need to develop more centralized resources.

The importance of correlating multiple kinds of data favors sample preparation performed at a single facility. Ideally, single identifiable samples should be used for measurements in many different categories. In addition, the same person at the same facility should perform a single category of measurement with different samples. These criteria lead to two organizational models for experimentation. In one model, sample preparation and all the measurements are to be performed at a single large experimentation center. The Max Planck Institute for Molecular Plant Physiology in Germany, RIKEN Yokohama Institute in Japan, and the Institute for Advanced Biosciences in Japan are examples of such large centers. The other model is a consortium of several facilities, each of which is exclusively specialized, e.g., one site for all the sample preparation and other sites for each category of measurement. This second example is not common and the panel saw few examples, except perhaps for the Hepatocyte Project in Germany and the Alliance for Cell Signaling in the U.S. The large center model has advantages in better communication among the involved members, a lower chance of mistake/accident in experiment/data tracking, and higher cost-efficiency in operation. The large center model can also offer an opportunity for better communication between researchers in experimental and theoretical work by having such people at the same site. As emphasized in the beginning of the chapter, close interactions between experimental and theoretical work is crucial in success of systems biology research. The large center model has a disadvantage in requiring larger initial investment and with less flexibility as an organization in the long term. It can, however, accommodate focused research efforts by having programs for *ad hoc* experimentation teams. The consortium model can do this easily by adding appropriate laboratories as consortium members. In either model, operations at each site need to be tightly controlled for high-quality data generation. This is where top-down organizations work better. Such a center or consortium would be a major data generation site for a biological system, and, therefore, it is reasonable for it to take a lead in organizing "the" database for the particular biological system.

Even if we do not choose as extreme an option as a large center or consortium, top-down organizational efforts will become more important in data generation and management because of the need for standardization and easy access to investigators. This will impact research communities from a social viewpoint. In top-down organizations, the role of an individual becomes more team-oriented, and it will be more difficult to single out accomplishments made by the individual. This situation is not very compatible with current academic evaluation criteria for merit. We will need to establish a different set of criteria or a different career path, so that team-oriented researchers can develop their careers. In top-down organizations, leaders also need to have strong management skills, which are usually not taught during typical scientific training. Furthermore, to make the situation fair to researchers not involved in the center/consortium, rapid dissemination of data generated by the center/consortium should be enforced. Although rapid data dissemination in genome sequencing projects has been the norm, rapid data dissemination in large experiments that involve sophistication in designing and performing experiments is not yet very common. To ensure that this and other functions are optimally developed for the benefit of science, research communities will need to be involved in governance.

The future of data and database aspects in systems biology research lies more on cooperation than competition. This is necessary to effectively utilize limited funding and human resources. Cooperation among funding agencies at inter-program, inter-agency, and international levels will also be important to facilitate cooperation among researchers. The spirit of cooperation should be extended beyond academia and governments to industry. The panel saw much more involvement of industry in systems biology research in Japan and Europe (particularly in Japan) than in the U.S. The impression the panel got was that legal situations around intellectual properties may be different in Japan and Europe. Although the panel did not have a chance to closely study such legal issues, they are crucial in increasing the involvement of industry in the U.S.

REFERENCES

Alon, U., Surette, M. G., Barkai, N., and Leibler, S. 1999. Robustness in bacterial chemotaxis. *Nature* 397: 168–171.

Bar-Joseph, Z., Gerber, G. K., Lee, T. I., Rinaldi, N. J., Yoo, J. Y., Robert, F., Gordon, D. B., Fraenkel, E., Jaakkola, T. S., Young, R. A., and Gifford, D. K. 2003. Computational discovery of gene modules and regulatory networks. *Nat Biotechnol* 21: 1337–1342.

Bolstad, B. M., Irizarry, R. A., Astrand, M., and Speed, T. P. 2003. A comparison of normalization methods for high density oligonucleotide array data based on variance and bias. *Bioinformatics* 19: 185–193.

Carlson, J. M., and Doyle, J. 2002. Complexity and robustness. *Proc Natl Acad Sci U S A* 99 Suppl 1: 2538–2545.

Gunawan, R., Cao, Y., Petzold, L., and Doyle, F. J., 3rd. 2005. Sensitivity analysis of discrete stochastic systems. *Biophys J* 88: 2530–2540.

Hastie, T., and Tibshirani, R. 2004. Efficient quadratic regularization for expression arrays. *Biostatistics* 5: 329–340.

Hirai, M. Y., Klein, M., Fujikawa, Y., Yano, M., Goodenowe, D. B., Yamazaki, Y., Kanaya, S., Nakamura, Y., Kitayama, M., Suzuki, H., *et al.* 2005. Elucidation of gene-to-gene and metabolite-to-gene networks in Arabidopsis by integration of metabolomics and transcriptomics. *J Biol Chem* 280: 25590–25595.

Kitano, H. 2002. Systems biology: a brief overview. *Science* 295: 1662–1664.

Lanckriet, G. R., De Bie, T., Cristianini, N., Jordan, M. I., and Noble, W. S. 2004. A statistical framework for genomic data fusion. *Bioinformatics* 20: 2626–2635.

Lee, E., Salic, A., Kruger, R., Heinrich, R., and Kirschner, M. W. 2003. The roles of APC and Axin derived from experimental and theoretical analysis of the Wnt pathway. *PLoS Biol* 1: E10.

Mori, H. 2004. From the sequence to cell modeling: comprehensive functional genomics in Escherichia coli. *J Biochem Mol Biol* 37: 83–92.

Rao, C. V., Frenklach, M., and Arkin, A. P. 2004. An allosteric model for transmembrane signaling in bacterial chemotaxis. *J Mol Biol* 343: 291–303.

Segal, E., Shapira, M., Regev, A., Pe'er, D., Botstein, D., Koller, D., and Friedman, N. 2003. Module networks: identifying regulatory modules and their condition-specific regulators from gene expression data. *Nat Genet* 34: 166–176.

Ueda, H. R., Hayashi, S., Chen, W., Sano, M., Machida, M., Shigeyoshi, Y., Iino, M., and Hashimoto, S. 2005. System-level identification of transcriptional circuits underlying mammalian circadian clocks. *Nat Genet* 37: 187–192.

Wiley, H. S., Shvartsman, S. Y., and Lauffenburger, D. A. 2003. Computational modeling of the EGF-receptor system: a paradigm for systems biology. *Trends Cell Biol* 13: 43–50.

CHAPTER 3

Network Inference

Frank Doyle and Douglas Lauffenburger

INTRODUCTION

As systems biology emerges in the post-genomic era, the emphasis is shifting from annotation of individual genes and gene products to ascertaining how DNA-protein and protein-protein interactions occur within a complex network of structural, metabolic, and regulatory pathways in cells. This goal is, of course, aligned with that pursued in "reductionist" molecular cell biology for the past two decades, in which efforts to identify and characterize pathways typically have proceeded in a component-centric manner, beginning with an initial gene or protein of particular interest and attempting to ascertain other genes and/or proteins involved in the same pathway. However, although component-centric approaches have been successful in assembling most of the available knowledge about pathways to date, they have several inherent difficulties. First is the time required: accurate models of pathway function emerge only after evidence is accumulated over many years, with the work of many researchers at many laboratories. Second, these approaches do not directly reveal how multiple pathways influence each other or reveal this crosstalk only accidentally. Third, the vast amount of information on the various intracellular pathways remains fairly decentralized, buried across primary literature or within narrowly defined reviews.

Systems biology can offer an accelerated approach to this goal of identifying networks by which genes and proteins interact to carry out cellular operational and regulatory functions, using computational mining methods on high-throughput experimental data. The resulting high-level models are finding increasing utilization as tools for drug discovery, both by small companies as well as large pharmaceutical companies such as Eli Lilly and

M. Cassman et al. (eds.), Systems Biology, 31–45.

Novartis (Henry, 2005). In general, the resulting "high-level," topological models can then serve as a foundational prelude, if one wishes, for more familiar (to engineers, physicists, and applied mathematicians, at least) "low-level," mechanistic models. This categorization has previously been laid out (Ideker & Lauffenburger 2003), and the summary and Fig. 3.1 below are largely taken from their discussion by a common co-author of this chapter and that article.

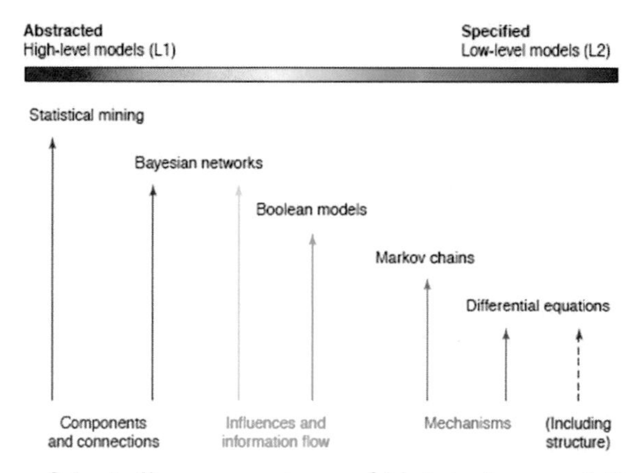

Figure 3.1. A diverse spectrum of high-to-low modeling approaches (Ideker and Lauffenburger, 2003).

Signaling and regulatory pathways consist of some number of components—such as genes, proteins and small molecules—wired together in a complex network of intermolecular interactions. Recent tec-hnological developments are enabling us to define and interrogate these pathways more directly and systematically than ever before, using two complementary approaches. First, it is now possible to systematically measure the molecular interactions themselves, by screening for protein-protein, protein-DNA and small molecular interactions. Several methods are available for measuring protein-protein interactions on a large scale—two of the most popular being the yeast two-hybrid system and protein co-immunoprecipitation (coIP) in conjunction with tandem mass spec-trometry. Although the vast majority of protein interactions have been generated for the budding yeast *Saccharomyces cerevisiae*, protein interactions are becoming available for a variety of other species including *Helicobacter pylori* and *Caenorhabditis elegans* and are catalogued in public databases such as BIND and DIP™. A current drawback of these high-throughput measurements is an associated high error rate. One approach for addressing this problem may be to integrate several complementary data sets together (e.g., two-hybrid interactions with coIP data or gene expression profiles) to reinforce the common signal.

Protein-DNA interactions, as commonly occur between transcription factors and their DNA binding sites, constitute another interaction type that can now be measured at high throughput. The relatively new technique of chromatin immunoprecipitation microarraying (so-called "ChIP-chip") has been used to characterize the complete set of promoter regions bound under nominal conditions for each of the >100 transcription factors in yeast, yielding >5,000 novel protein-DNA interactions in that organism. Additional types of pathway interactions, such as those between proteins and small molecules (carbohydrates, lipids, drugs, hormones and other metabolites), are difficult to measure on a large scale, although protein array technology might enable high-throughput measurement of protein-small molecule interactions in the near future.

In addition to characterizing molecular interactions, a second major approach for interrogating pathways is to systematically measure the molecular and cellular states induced by the interaction wiring. For example, global changes in gene expression are measured with DNA microarrays, whereas changes in protein abundance, protein phosphorylation state, and metabolite concentrations can be quantified with mass spectrometry, Nuclear magnetic resonance (NMR) and other advanced techniques. Of these approaches, measurements made by DNA microarrays are currently the most comprehensive (every mRNA species is detected); high-throughput (a single technician can assay multiple conditions per week); well characterized (experimental error is appreciable, but understood); and cost-effective (whole-genome microarrays are purchased commercially for US $50 to $1000, depending on the organism). However, continued advances in protein labeling and separation technology are making the measurement of protein abundance and phosphorylation state almost as feasible, with the primary barrier being the expense and expertise required to set up and manage a mass spectrometry facility. Measurement of metabolite concentrations, an endeavor otherwise known as metabonomics, is currently limited not by detection (thousands of peaks, each representing a different molecular species, are found in a typical NMR spectrum) but by identification (matching each peak with a chemical structure is difficult). Clearly, measuring changes in cellular state at the protein and metabolic levels will be crucial if researchers are to gain insight into not only regulatory pathways, but also those pertaining to the cell's signaling and metabolic circuitry.

To arrive at a high-level topological model of a cellular network of interest, data on molecular interactions and states can be integrated in a multi-tiered strategy. First, the global molecular interaction scaffold is constructed from systematic measurements of protein-protein interactions, protein-DNA interactions and/or metabolic reactions (as detailed in the previous section). In the case of budding yeast, a minimal set might include 14,941 protein-protein interactions catalogued in the DIP™ database; 5,631 protein-DNA interactions from a combination of TRANSFAC® and ChIP-chip; and 599 enzymatic reactions in MetaCyc. Second, this scaffold is filtered against changes in mRNA expression, protein expression and/or post-translational modifications recorded in response to different cellular perturbations. Networks within the interaction scaffold with mRNA or protein states that are significantly activated by perturbation are identified and mapped according to a computational search engine. The interaction pathways and complexes comprising the scaffold constitute topological models, which are then prime candidates for further verification and modeling as important signaling and compensatory mechanisms controlling the cellular perturbation response. The key advance of these searches is that by integrating two complementary global data sets, it is possible to condense and partition the enormous quantity of data into a small number of relevant pieces suitable for lower level modeling.

Examples of this general scheme have been reported in recent literature. Several groups have applied "co-clustering" approaches to identify groups of proteins that are both differentially expressed under similar sets of conditions and closely connected by the same network of interactions in the scaffold. In many cases, these "expression-activated networks" correspond to well known protein complexes, regulatory pathways or metabolic reaction pathways. Other groups have used probabilistic approaches to match changes in gene expression with the transcription factors that are most likely to regulate them directly. These methods start with a cluster of differentially expressed genes and incrementally choose a small set of transcription factors that, by virtue of their levels and/or interactions in the scaffold, can maximally predict the observed levels of differential expression in the cluster. New transcription factors are added only if they lead to a sufficient increase in predictive power over the transcription factors already in the model.

Several software tools are now available for visualizing interaction scaffolds [Osprey, http://biodata.mshri.on.ca; PIMRider®, http://pim. hybrigenics. com; GenoMax™, http://www.informaxinc.com; Cytoscape, http://www.cytoscape.org; Pathway Tools, http://bioinformatics.ai.sri.com/ptools/]. For instance, the Cytoscape framework provides network visualization, layout and annotation, as well as clustering of the network against expression data to generate topological network models. The Pathway-Tools component of the MetaCyc metabolic pathway database can superimpose enzyme expression levels on the map of biochemical reactions for a species, giving a good indication of which reaction pathways are most affected over a panel of growth conditions profiled by microarray.

Because DNA microarray technology is currently much more widespread than technologies for protein or metabolite profiling, the vast majority of these approaches have used gene expression profiling as the primary state measurement. Of course, pathway mapping methods based on mRNA profiling alone capture just one facet of a much larger and complex cellular response. As it becomes possible to measure cellular state at the protein and small-molecule level, researchers expect that algorithms similar to those described above will emerge. Currently, omitting this information from the analysis means that regulatory networks not purely transcriptional in nature remained to be elucidated in this high-throughput computation-aided manner. Indeed, the next obvious challenge beckoning is to push forward from the current focus on transcriptional regulation to regulation occurring in the post-transcriptional and post-translational arenas, and especially arising from extracellular cues.

NETWORK INFERENCE AND MODEL STRUCTURE

In the area of network inference, the models are primarily static interconnection descriptions of collections of proteins, metabolites, and/or genes. The "inference" problem involves the estimation of the interactions of elements in the network, given (possibly time series) data of activities of different nodes (e.g., gene interactions from gene expression data). The goals of the inference problem are multiple, and include: (i) hypothesis generation, (ii) design of experiment, (iii) understanding of cellular function, and (iv) unraveling design principles, among others. The sources of information for these inference problems include large-scale deletion projects, and vast numbers of microarray experiments. In the early years of bioinformatics studies, the structural localization properties were inferred (e.g., which transcription factors regulate the transcription of which genes), although experimental methods now exist for identifying protein-DNA interactions on a genomic scale, such as ChiP assays, that yield structural knowledge.

Given the wide variety of modeling objectives, as well as the heterogeneous sources of data, it is not surprising that the WTEC study team observed many approaches to modeling for network inference in Japan, Europe, and the U.S. For the case of microbial systems biology, the review paper by Stelling (2005) provides a good summary of the spectrum of modeling approaches. He classifies the modeling efforts in two respects: network complexity and level of detail. This chapter of the report focuses on the problems associated with the more complex, less detailed models, while a separate chapter examines the issues associated with more detailed (mechanistic) models.

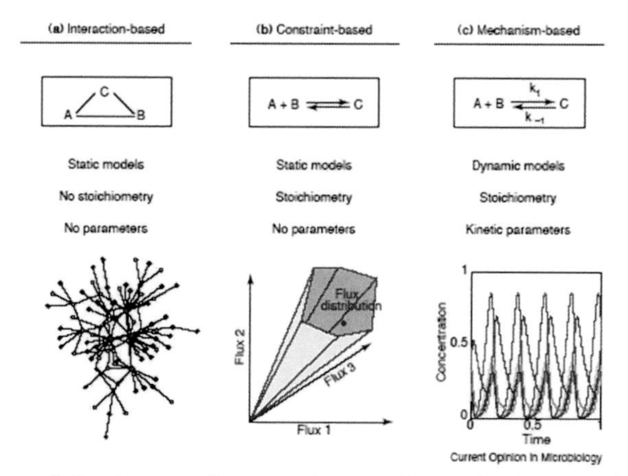

Figure 3.2. Approaches to the mathematical modeling of cellular networks (Stelling, 2005).

The mathematical structures invoked to capture network interactions are numerous, and include:

- *Boolean Networks*—in which the network is represented as a graph of nodes, with directed edges between nodes and a function for each node (e.g., Ideker et al., 2000)
- *Petri Nets*—another graph theoretic structure in which nodes (or places) are connected by arcs and activities are modeled by transitions (e.g., Nagasaki et al., 2004)
- *Bayesian Nets*—combine directed acyclic graphs with a conditional distribution for each random variable (vertices in graph) (e.g., Pe'er et al., 2001)
- *Signed Directed Graphs*—another graph theoretic structure in which a signed directed edge is used to represent activation versus inhibition (depending on sign) (e.g., Kyoda et al., 2004)
- *S-systems*—notably a *dynamic* approach in which polynomial nonlinear dynamic nodes are used to capture network behavior (e.g., Kimura et al., 2005)

A significant challenge in constructing these network models from data, particularly for gene network models, is the fact that the node dimension (number of genes) can be on the order of 10,000—leading to a computationally untenable problem for inference (i.e., determination of 10^8 coefficients of interaction!). In reality the network is tremendously sparse and highly structured, such that there are orders of magnitude fewer "interactions" that must be captured with coefficients. The knowledge that not every gene regulates every other gene, and the fact that not every transcription factor regulates every gene can be exploited to prune significantly the number of coefficients for network identification.

A related concept that can be exploited is the knowledge that the low dimensional connection structures in these networks obey regular hierarchies, which create opportunities for structured model identification. Many biophysical networks can be decomposed into modular components that recur across and within given organisms. One hierarchical classification is to label the top level as a *network*, which is comprised of interacting regulatory *motifs* consisting of groups of 2–4 genes (Lee et al., 2002; Shen-Orr et al., 2002; Zak et al., 2003). At the lowest level in this hierarchy is the *module* that describes transcriptional regulation, of which a nice example is given in Barkai and Leibler, 2000. At the *motif* level, one can use pattern searching techniques to determine the frequency of occurrence of these simple motifs (Shen-Orr et al., 2002), leading to the postulation that these are basic building blocks in biological networks. Many of these components have direct analogs in system engineering architectures. Consider the three dominant network motifs found in *E. coli* (Shen-Orr et al., 2002): (i) feed-forward loop, (ii) single input module, and (iii) densely overlapping regulon. Similar studies in a completely different organism, *S. cerevisiae*, yielded six related or overlapping network motifs (Lee et al., 2002): (i) autoregulatory motif, (ii) feed-forward loop, (iii) multi-component loop, (iv) regulator chain, (v) single input module, and (vi) multi-input module.

Beyond structural classification, one can analyze these motifs for their functional character, as shown by Wolf and Arkin, 2003, and again, one finds the recurring *dynamic* functional motifs in circuits and signal processing: (i) switches, (ii) oscillators, (iii) amplitude filters, (iv) bandpass filters, (v) memory, (vi) noise filters, and (viii) noise amplifiers.

In effect, these studies demonstrate that, in both eukaryotic and prokaryotic systems, cell function is controlled by sophisticated networks of control loops that are cascading onto, and interconnected with, other (transcriptional) control loops. The noteworthy insight here is that the complex networks that underlie biological regulation appear to be constructed of elementary *systems* components, not unlike a digital circuit. This creates opportunity for the network inference methods that incorporate such knowledge via constrained search methods, or exploiting prior knowledge in Bayesian frameworks.

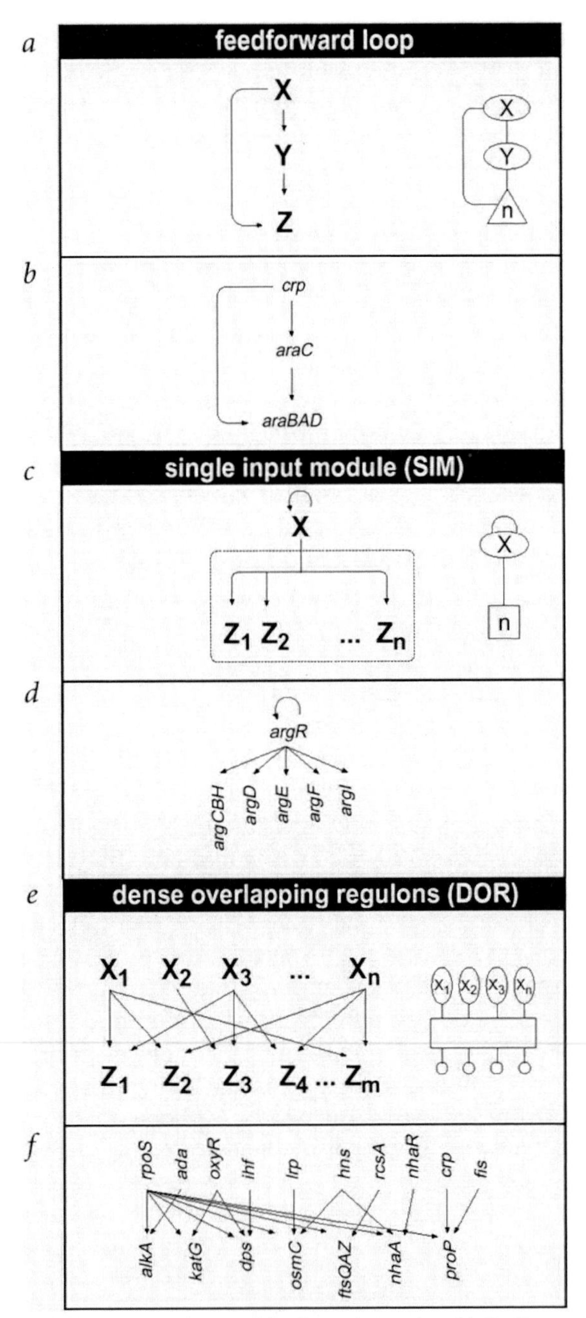

Figure 3.3. Network motifs found in the E.Coli transcriptional regulation network.

In addition to the two classes of models mentioned previously (based on complexity and detail), there is an intermediate class consisting of optimization-based models. In many respects, this class has a hybrid character of empiricism and fundamental details. The underlying assumption is that cells have been organized over evolutionary time scales to optimize their operations in a manner consistent with mathematical principles of optimality. The cybernetic approach developed by Ramkrishna and co-workers (Varner and Ramkrishna, 1998) is founded on a simple principle; evolution has programmed or conditioned biological systems to optimally achieve physiological objectives. This straightforward concept can be translated into a set of optimal resource allocation problems that are solved at every time step in parallel with the model mass balances (basic metabolic network model). Thus, at every instant in time, gene expression and enzyme activity is rationalized as choice between sets of competing alternatives each with a relative cost and benefit for the organism. Mathematically, this can be translated into an instantaneous objective function. The potential shortcoming is a limited handling of more flexible objective functions that are commonly observed in biological systems. An alternative approach is the Flux Balance Analysis (FBA) (Watson, 1986), in which a suitable linear programming problem is posed and solved (Edwards et al., 1999). The resulting model is not a dynamic model, and does not yield an analytical formulation, but the computational solution time is modest, and the approach has yielded success for a number of biological examples. Essential to the development of the model are the formulation of the system constraints, consisting of: (i) stoichiometric constraints that represent flux balances; (ii) thermodynamic constraints to restrict the directional flow through enzymatic reactions; and (iii) physicochemical capacity constraints to account for maximum flux through individual reactions. Recent extensions have addressed the problem of regulation by including additional time-dependent constraints in the formulation. The incorporation of transcriptional regulatory events in the FBA framework has extended the validity of the methodology for a number of complex dynamic system responses (Covert et al., 2001). In an alternate formulation, dynamic mechanistic details are incorporated as constraints leading to a dynamic FBA extension (Mahadevan et al., 2002).

As noted several times in this report, *dynamic* behavior is an essential property of complex biophysical networks that must be captured in models of those networks. There are some preliminary ideas in capturing network behavior in the form of dynamic models—both discrete time (Hartemink et al., 2002) and continuous (Zak et al., 2004). Many challenges exist in developing dynamic models from the type of data that is typically generated in the corresponding experiments, including: (i) sampling rate is rarely uniform and (ii) data is often combined with other labs, introducing a number of biases. The previously noted problems of the curse of dimensionality are more pronounced in the case of dynamic models, if one augments the network interconnection dimensionality with a large number of possible dynamic states (activated, repressed, silenced, etc.), let alone the full continuum of dynamic response.

VALIDATION, ITERATION, DISCRIMINATION, AND IDENTIFIABILITY

One of the major issues in the reverse engineering of a genetic regulatory network is the challenge of *uniquely* identifying the gene interactions (i.e., model parameters) from experimental data, such as gene expression profiling. This issue, known as identifiability in control theory (Ljung, 1999), deals with the information content of the data; the quantity and quality of the measurements with respect to the model parameters. Recent work in the U.S. and in Europe on the identifiability of gene networks revealed that full knowledge of gene interconnections and perfect measurements still could not guarantee full identifiability of gene interactions (Zak et al., 2003), and, furthermore, that improved experi-mental protocol was far more effective than increased measurements (J. Stelling, unpublished data, 2005). The latter study points to the fact that perturbations should be designed strategically. Typical knockouts involve so-called "direct effects" in which the expression level of various genes is altered in a network arrangement that involves direct connectivity to cis-regulatory elements of downstream genes (possible multiple cascades). An "indirect effect" can also be used in which a mediating component (e.g., mRNA) is introduced to correct an intermediate element in the direct action cascade described previously.

Coupled to this, noise in the measurements and the inherent stochastic nature of gene expression make practical identification of genetic regulatory networks difficult. In practice, the reverse engineering of a gene network should involve a careful design of the experiments using prior knowledge of the system, to obtain the most informative measurements. Further, this process should be iterative in which the result from each trial is used to better design the next experiment. Here, a measure of informativeness of data, such as the Fisher Information Matrix (FIM), can lead to a formal procedure for the optimal design of the experiment. Aside from the aspect of the quality of data, another practical limitation in most (if not all) of the reverse-engineering of a gene network is the limited quantity of data, in terms of sampling frequency and number of independent measurements. For example, although gene expression profiling can provide high-throughput data to estimate interactions among thousand of genes, this method still does not depict the protein-mediated regulatory effects. In many cases, parameter estimation from limited measurements suffers from stringent computational requirement and degeneracy, where many parameter combinations give similar agreement to the observed behavior. Here, measurement selection procedures can help identify the combination of measurements that give the best identifiability.

Given the iterative nature of this framework for model development and refinement of experimental protocols, a termination criterion must be established. In the application domain of systems engineering, it is understood that for certain experimental data, it is not possible to confirm whether the model is really valid; however, one can conclude whether the model is not contradicted by the given data (Poolla et al., 1994). Such model (in)validation tests can be formulated for the network inference problems described in this chapter, and are usually based on the difference between the simulated and measured output and some statistics about these differences. Typical statistics for the model errors include maximum absolute value, mean value and variance. These methods are slowly migrating from the engineering domain and are likely to find greater application in systems biology as experimental methods are refined, and closer collaborations are developed between modelers and experim-entalists.

REGRESSION

As described above, most of the network inference work undertaken to date has been aimed at elucidating relationships among components in a network. However, the operation of a network is generally important only within the context of a physiological function it carries out or regulates at the level of individual cells or cell populations (e.g., tissues). Thus, it is essential to consider efforts to elucidate relationships between network components (or, more appropriately, network component properties such as levels, states, locations, and/or activities) and downstream cellular behavioral functions. Computational models for these kinds of "signal-response" relationships are much more difficult to formulate than for the more commonly-studied "cue-signal" relationships by which network component properties are governed by extracellular (or intracellular) stimuli, because the biochemical and biophysical processes involved in the former are much less well understood as well as less proximal. Nonetheless, several research efforts are underway that take as their objective the development of inferring "signal-response" relationships for network component regulation of cell functions. One class of methods which appear to be useful for inferring dependence of cell functions on network component properties are those founded on principal components analysis (Janes, 2004), including a variant termed network components analysis (Kao, 2004). This class of methods permits determination of the most critical combinations of network component properties for correlation, and even prediction, of functional responses. A second class of methods also being employed for this purpose is decision tree analysis (Hautaniemi, 2005). An advantage of decision tree analysis is that the combinations of network component properties associated with functional responses are explicitly delineated in direct manner, permitting the "logic" of how the network component properties combine to govern functional behavior to be viewed and interpreted easily. Finally, Bayesian network analysis has been used for a similar purpose (Woolf, 2005). Here, the "logic" connecting network component properties to one another and to the functional behavioral response is more complex to interpret, but is nonetheless available.

COMPARISON OF EFFORTS IN EUROPE, JAPAN, AND THE U.S.

Implicit in the preceding sections was a comparative analysis of efforts in the U.S., Japan, and Europe (as well as other regions such as Israel) by virtue of the cited references. For the purposes of the study, we outline from additional specific highlights in this area of network inference. It is worth noting that many of the ideas described in this chapter fall into the area of *Bioinformatics*, which has arguably gained a foothold in all of the geographical regions considered, with consideration to research, education, and infrastructure.

In the region of Japan, there were numerous significant advances in the area of network inference. The Kitano Laboratory (Symbiotic Systems Project) plays a leading role in the development of the Systems Biology Markup Language (SBML) and the establishment of standards in modeling biological systems. In addition, they are conducting research in the area of regression algorithms for network inference. The RIKEN Yokohama Institute was, as noted elsewhere, an extremely large-scale operation, and there were numerous laboratories addressing important problems in network inference, including: (i) inference algorithms (cooperatively coevolutionary), (ii) the formation of a consortium for the study of receptor tyrosine kinase regulatory networks, and (iii) the dynamic profiling of regulation in circadian networks by the Ueda Laboratory (Kobe). The Miyano Laboratory (U. Tokyo) was also developing regression algorithms for network inference of gene regulatory networks in yeast, and notably, they are conducting drug development studies with pharmaceutical companies. The Computational Biology Research Center (Tokyo) was conducting network inference studies for application to lung cancer. The Kanehisa Laboratory (Kyoto University), well known for the Kyoto encyclopedia of genes and genomes (KEGG) database, is conducting research in the reconstruction of dynamic networks via kernel methods, and is also enabling portability of the KEGG database networks to SBML and genomic object net (GON) models formats. This can be useful for the ultimate development of mechanistic large-scale network models. Finally, the Ito Laboratory at the University of Tokyo was utilizing heterogeneous measurements for network inference (MS, FRET, ChiP, GATC-PCR, etc.), which is viewed as essential to overcome the identifiability issues described earlier in this chapter.

In Europe, the Max Planck Institute for Dynamics of Complex Systems (headed by Prof. Gilles) was one of the few groups that was explicitly addressing the challenges in identifiability, model iterations, perturbations, and design of experiment. At Humboldt University in Berlin, there were efforts described for dynamic modeling from microarray data, with application to the Ras pathway. The Reuss Laboratory (Stuttgart) described bioinformatics studies with application to cytochrome p450. The Armitage Laboratory (Oxford) emphasized bottom-up approaches for pathway analysis in histidine sensing. The Noble Laboratory (Oxford) challenged the strict bottom-up and top-down approaches, advocating a combination that starts in the middle.

A number of groups are vigorously active in the United States in the area of network inference, across all the technical areas outlined above. The Ideker group at UC-San Diego and the Ingber group at Harvard Medical School are pursuing Boolean approaches. The Gifford group at MIT and the Koller group at Stanford are employing Bayesian network approaches for investigation of networks focused on genomic data, and by the Lauffenburger group at MIT for investigation of networks focused on proteomic data. Regression methods, such as principal components analysis, network components analysis, and decision tree analysis are being used by the Liao group at UC-Los Angeles to study genomic networks and their relationship to physiological functions, and by the Lauffenburger group at MIT to study proteomic networks and their relationship to physiological functions. Cybernetic approaches are championed by the Ramkrishna group at Purdue, and flux balance methods—mainly for metabolic networks—by the Palsson group at UC-San Diego and the Stephanopoulos group at MIT.

SUMMARY

In summary, one finds numerous network inference studies in all of the regions described with the U.S., Japanese, and Israeli groups leading in the development of methodologies. All regions showed exciting application studies, with significant potential for "success stories" to emerge in the coming years.

The encouraging trends that were observed included: (i) multiple, complementary approaches to the regression of models for network inference,

(ii) motifs and modules being incorporated into network inference methods, (iii) a nice interplay emerging between the classical static network databases and the formats for dynamic systems biology models (*e.g.*, SBML), and (iv) a considerable amount of curricular development in this area (notably in bioinformatics).

Of concern was the fact that the issues of: (i) explicit incorporation of dynamics, (ii) identifiability and (in)validation of models, and (iii) model iterations with design of experiment, were receiving only modest attention in the regions, with noteworthy efforts in the U.S. and Europe (particularly Germany). There were many reported examples of researchers identifying large numbers of parameters from relatively small data sets. However, there appear to be a number of groups working towards solutions to these challenges, and considerable progress can be expected in the next two to three years.

REFERENCES

Barkai, N. and S. Leibler. 2000. Circadian Clocks Limited by Noise. *Nature* 403: 267–8.

Covert, M. W., C. H. Schilling and B. Palsson. 2001. Regulation of gene expression in flux balance models of metabolism. *J Theor Biol* 213: 73–88.

Edwards J. S., R. Ramakrishna, C. H. Schilling and B. O. Palsson. 1999. Metabolic Flux Balance Analysis. In: *Metabolic Engineering* (S. Y. Lee and E. T. Papoutsakis, Eds.). pp. 13–57. Marcel Deker.

Hartemink A. J., D. K. Gifford, T. S. Jaakola, R. A. Young. 2002. Combining Location and Expression Data for Principled Discovery of Genetic Regulatory Network Models. *Proc Pac Symp Biocomput* 7: 437–449.

Hautaniemi S., S. Kharait, A. Iwabu, A. Wells, and D. A. Lauffenburger. 2005. Modeling of signal-response cascades using decision tree analysis. *Bioinformatics*, in press.

Henry C. M. 2005. Systems Biology: Measurement and Modeling Approaches Bring a Big-Picture View of Biology and May Improve Drug Discovery and Development. *Chem & Eng News* 83(7): 47–55.

Ideker T. E., V. Thorsson, and R. M. Karp. 2000. Discovery of Regulatory Interactions Through Perturbation: Inference and Experimental Design. *Pac Symp BioComputing* 5:302–313.

Janes K. A., J. R. Kelly, S. Gaudet, J. G. Albeck, P. K. Sorger, and D. A. Lauffenburger. 2004. Cue-signal-response analysis of TNF-induced apoptosis by partial least squares regression of dynamic multivariate data. *J Comput Biol* 11: 544–561.

Kao K. C., Y. L. Yang, R. Boscolo, C. Sabatti, V. Roychowdhury, and J. C. Liao. 2004. Transcriptome-based determination of multiple transcription regulator activities in *Escherichia coli* by using network component analysis. *Proc Natl Acad Sci U S A* 101: 641–646.

Kimura S., K. Ide, A. Kashihara, M. Kano, M. Hatakeyama, R. Masui, N. Nakagawa, S. Yokoyama, S. Kuramitsu, and A. Konagaya. 2005. Inference of S-system Models of Genetic Networks using a Cooperative Coevolutionary Algorithm. *Bioinformatics* 21: 1154–1163.

Kyoda K., K. Baba, S. Onami, and H, Kitano. 2004. DBRF-MEGN Method: An Algorithm for Deducing Minimum Equivalent Gene Networks from Large-Scale Gene Expression Profiles of Gene Deletion Mutants. *Bioinformatics* 20: 2662–2675.

Lee T. I., N. J. Rinaldi, F. Robert, D. T. Odom, Z. bar-Jospeh, G. K. Gerber, N. M. Hannett, C. T. Harbison, C. M. Thompson, I. Simon, J. Zeitlinger, E. G. Jennings, H. L. Murray, D. B. Gordon, B. Ren, J. J. Wyrick, J. B. Tagne, T. L. Volkert, E. Fraenkel, D. K. Gifford and R. A. Young. 2002. Transcriptional Regulatory Networks in Saccharomyces cerevisiae. *Science* 298: 799–804.

Ljung L. 1999. System Identification: Theory for the User, Englewood Cliffs, NJ, Prentice Hall.

Mahadevan R., J. Edwards and F. J. Doyle III. 2002. Dynamic flux balance analysis of diauxic growth in *Escherichia coli. Biophys J* 83: 1331–1340.

Nagasaki M., A. Doi, H. Matsuno, and S. Miyano. 2004. A Versatile Petri Net Based Architecture for Modeling and Simulation of Complex Biological Processes. *Genome Informatics* 15: 180–197.

Pe'er D., A. Regev, G. Elidan, and N. Friedman. 2001. Inferring Subnetworks from Perturbed Expression Profiles. *Bioinformatics* 17: S215-S224.

Poolla K., P. Khargonekar, A. Tikku, J. Krause, and K. Nagpal. 1994. A Time-Domain Approach to Model Validation. *IEEE Trans Automat Contr* 39: 951–959.

Shen-Orr S. S., R. Milo, S. Mangan, and U. Alon. 2002. Network Motifs in the Transcriptional Regulation Network of *Escherichia coli. Nature Genetics* 31: 64–68.

Stelling J. 2004. Mathematical Issues in Systems Biology. *Current Opinion in Microbiology* 7: 513–518.

Varner J. and D. Ramkrishna. 1998. Application of cybernetic models to metabolic engineering: Investigation of storage pathways. *Biotech Bioeng* 58: 282–291.

Watson M. R., 1986. A discrete model of bacterial metabolism. *Comp Appl Biosciences* 2: 23–27.

Wolf D. M. and A. P. Arkin. 2003. Motifs, Modules, and Games in Bacteria. *Current Opinion in Microbiology* 6: 125–134.

Woolf P. J., W. Prudhomme, L. Daheron, G. Q. Daley, and D. A. Lauffenburger. 2005. Bayesian analysis of signaling networks governing embryonic stem cell fate decisions. *Bioinformatics* 21: 741–753.

Zak D., G. Gonye, J. S. Schwaber and F. J. Doyle III. 2003. Importance of Input Perturbations and Stochastic Gene Expression in the Reverse Engineering of Genetic Regulatory Networks: Insights from an Identifiability Analysis of an *In Silico* Network. *Genome Research* 13: 2396–2405.

Zak D., R. K. Pearson, R. Vadigepalli, G. Gonye, J. S. Schwaber and F. J. Doyle III. 2004. Continuous-time identification of gene expression models. *Omics* 7: 373–386.

CHAPTER 4

Modeling and Network Organization

Cynthia Stokes and Adam Arkin

INTRODUCTION

The use of mathematical modeling and analysis of networks has a long history in biological research. Perhaps the best-known early example of insightful modeling is the work of Hodgkin and Huxley in 1952 describing how sodium and potassium ion channels could function together to produce the membrane action potential in neurons (Hodgkin and Huxley, 1952). For several decades, models and theory were mostly the domain of applied mathematicians, physical scientists and engineers, many of whom worked rather independently of experimental science and the work remained somewhat obscure and theoretical. With the broad availability of computers and IT infrastructure that has emerged in the last several decades, the use of modeling and theory in biological research has expanded greatly, as has the size of the models being developed.

Historically, much modeling was used to study and interpret what could be directly observed in the laboratory, namely, functions of cells, tissues, organs and organism physiology. In addition, starting with the work of Jacob and Monod there was a great deal of biochemical and genetic modeling. Enzymological modeling such as that by Garfinkel was mechanistically very detailed and chemically supported. On the other hand, the genetic network models were far more abstract and rarely related to data. However, it was recognized that in many cases that there was a common underlying mathematical framework to both. Techniques from non-linear dynamics, chemical engineering analyses (stoichiometric network analysis, etc.) were brought to bear. More abstract models of these processes were used to study the possible organization and optimalities of biological

M. Cassman et al. (eds.), Systems Biology, 47–81.

networks. Jim Bailey, Goodwin, and others wrote monographs on the topic. And of course, Turing himself made the first links from chemistry to development.

With the advent of molecular biology and ensuing capabilities in genomics, proteomics, and so forth, mathematical models are now being used extensively to study intracellular molecular networks such as kinase cascades and metabolic pathways, as well as gene regulatory networks. These intracellular molecular networks are a primary subject of network organization analysis, along with epidemiological networks, and structural networks such as lung airway, vascular, and neural network topologies. At the same time, modeling of multicellular networks with multiple intercellular interactions, and sometimes multiple anatomical scales (biochemicals, cells, tissues, organs), has continued and in the last 5–10 years the biological breadth and detail of such models has increased dramatically. This has been possible because of both the increased availability of data that informs both model structure and parameter values, as well as the availability of sufficient computational power with which to code and solve them.

The growing popularity of modeling in biological research is evident from the increasing number of public forums dedicated to or including it. (Akutsu, Miyano et al., 2000) The number of research conferences including or fully devoted to biological modeling has increased dramatically in the last five years and are too numerous to list here. As noted in the introductory chapter, new journals devoted to systems biology, including modeling and network organization, have been launched, including *In Silico* Biology, PLoS Computational Biology, IEE Proceedings Systems Biology, and Molecular Systems Biology.

Applications of Models

Modeling and analysis are uniquely suited to a wide variety of applications. Models can be used to test specific hypotheses, for instance, about how a system is structured or how it functions. They can be used to make predictions, which can then be tested with appropriate experiments. Models can be used in a more exploratory manner, for instance, to discover the types of properties that might emerge from integration of parts with specific properties in specified ways. While experiments can also be used to test hypotheses, make predictions and explore, mathematical models explicitly represent components and their interactions in a controllable, manipulable environment (the computer) which allows calculation of how these things change through time and space (if those are included) and observation of every element and relationship in the model. The analogous experimental measurements would frequently be difficult or impossible and sometimes unethical. In addition to these specific applications in research, models are also excellent aids to communication and teaching. In contrast to static words or pictures on a page, they provide an interactive method for demonstrating and exploring how the modeled system works in an easy-to-use computer environment.

What is a Mathematical Model?

A model is a set of *structured assertions* that specify the *interactions* among *entities* of a *network*. An example of a model that illustrates this definition is given in Figure 4.1. The entities in a model can be many different things, such as properties of specific biological elements (e.g., molecular concentration, cellular density, or organ volume) or specific physiologic characteristics (e.g., blood pressure, heart rate, or weight). Likewise the interactions that are specified among the entities can represent various processes such as molecular reactions, binding of a molecule to a cell-surface receptor, subsequent stimulation of that cell, etc. From a model one can calculate how the entities change over time and/or through space or their value at steady state.

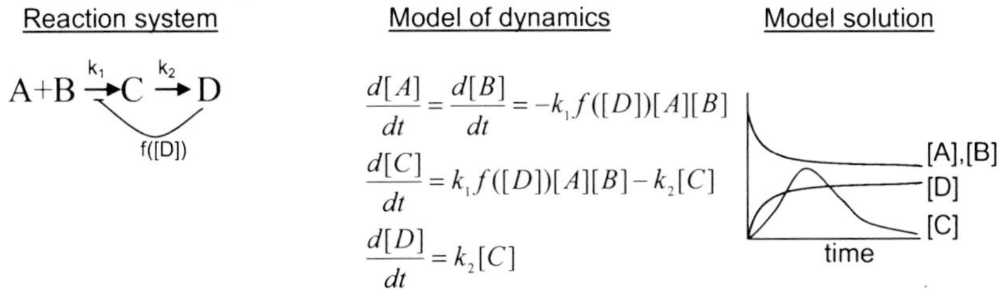

Figure 4.1. Schematic of a chemical reaction system where chemical species A and B react together to form C, which further changes into D, and D inhibits the first reaction. The equations of a mathematical model of that system using ordinary differential equations (ODEs) are shown along with a graph of a simulation of that model, assuming A and B start with equal, non-zero concentrations while C and D are initially zero. Brackets around a chemical name indicate concentration.

Mathematical models of biological systems can take many different forms, and the appropriate types of mathematical equations are highly dependent on the problem one is attempting to represent with the model. Many models utilize ordinary or partial differential equations to represent continuous, deterministic systems and partial differential equations if space or mechanics are involved. Various methods are used to include stochastic or probabilistic properties of the system including Langevin dynamics and the more physically rigorous chemical master equation. Discrete methods like particle/molecular dynamics, cellular automata and agent-based models are utilized where the actions of individual elements of a system, rather than the population behavior, is of interest.

What is Network Organization Analysis?

It is believed that biological networks are not randomly structured, but rather that various parts of the network have structures that provide specific functional units, and those structures can be found in multiple parts of the network. Although universal definitions are not agreed upon, these subunits with the same structure are often called motifs or modules, where the term motif is often used to indicate the smallest repeated unit and module to indicate a collection of units that form a "separable" functional group—that is, a group of processes that together create some well-defined behavior that is a pure input/output function and is not otherwise affected by inclusion of the network–although those definitions are not universal. For example, Segal et al. (2003) define a module in the context of gene expression as a group of genes that tend to respond in a joint manner, i.e., through temporally coordinated gene expression. Wuchty et al. (2003) define a module in the context of network topology as a discrete group of interconnected elements that is abstracted from the topology of the network. Von Dassow et al. (2000) define a module functionally as a set of genes and their products which, as an emergent consequence of their interactions, perform some task nearly autonomously.

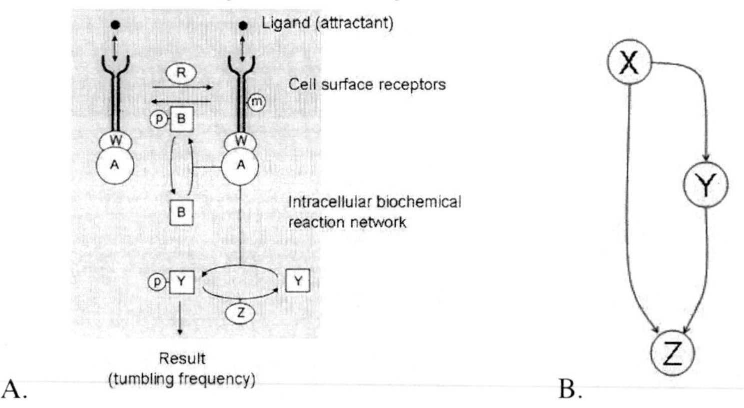

Figure 4.2. Example of (A) a molecular interaction network involved in bacterial chemotaxis adapted from Alon et al. (1999) and (B) a single motif, the feed-forward loop, frequently found within intracellular molecular networks.

Figure 4.2 illustrates molecular networks and motifs. These smaller units are thought to derive from functional need of the organism as well as fundamental physical principles (e.g., thermodynamics). The analysis and theory of network organization focuses on discovering underlying principles and motifs of systems and networks. Many different approaches to network organization analysis are being used in biology and various reviews are available (Alm and Arkin, 2003; Barabasi and Oltvai, 2004; Itzkovitz and Alon, 2005).

OVERVIEW OF WORLDWIDE EFFORTS

A wide variety of research efforts utilize biological modeling and network analysis throughout the U.S., Europe, Japan, and elsewhere. This report's summary of work in the different regions discusses modeling efforts in terms of several broad subjects, including intracellular gene regulatory and biochemical networks; cellular metabolism; receptor dynamics and cell function; multi-cellular/ tissue/ organ function; electrophysiology; organism development; and spatial organization and pattern formation. While somewhat arbitrary, these categories group problems that are frequently of common interest to a given set of researchers. Network organization research is discu-ssed as a separate category, and the efforts of industry in these areas are discussed in a separate section below.

Biological Modeling and Network Organization Analysis in the United States

The use of models in biological research in the U.S. is extensive; a comprehensive review is beyond the scope of this chapter. Instead, representative studies that illustrate the variety of ongoing research are highlighted and referenced.

The subject of signaling networks and gene regulatory networks has garnered much attention in recent years, and strong efforts at representing and understanding specific networks using models have been put forth by researchers such as Ravi Iyengar (Mt. Sinai School of Medicine), John Tyson (University of Virginia), Adam Arkin (University of California-Berkeley) Peter Sorger (Massachusetts Institute of Technology) and Hamid Bolouri (Institute for Systems Biology). Arkin, working with Harley McAdams, John Ross and recently Michael Samoilov, pioneered the study of the importance of stochastic chemical processes in gene expression and signal transduction. He demonstrated how, from physical chemical first principles, gene expression in prokaryotes is expected to show bursts and erratic production of proteins. In later work he showed that the fundamental fluctuations in reaction rates can result in qualitatively different behaviors than that predicted by standard mass-action kinetic models. He has followed up the implications of this noise in the lysis/lysogeny decision of λ-phage and type-1 pili phase variation among other systems and shown that the noise is a fundamental and biologically important part of the regulation of these systems. Van Oudenaarden and Elowitz have each separately followed-up this work with elegant theories and measurements of the effect in bacteria. Wolf and Arkin then recently showed under what conditions such non-genetic diversification mechanisms are part of an evolutionarily stable strategy.

Iyengar's research focuses on cellular signaling systems, utilizing close integration of experimental and theoretical methods, with emphasis on signaling associated with G-protein coupled receptors. For example, Iyengar and colleagues developed a model of the mitogen-activated protein kinase (MAPK)/protein kinase C (PKC) system, and in concert with targeted experimentation they demonstrated that this system can operate with one or two stable states and is therefore quite flexible in its ability to control cellular processes such as cell cycle (Bhalla, Ram, and Iyengar, 2002). Tyson and his collaborators have developed detailed models of the molecular control mechanisms of cell cycle in fission yeast, budding yeast, frog embryo and mammalian cells (Chen et al., 2004; Zwolak, Tyson, and Watson, 2005; Novak and Tyson, 2003; Sveiczer, Tyson, and Novak, 2004). They have worked closely with experimentalists to compare the function of the budding yeast cell cycle model in 131 mutant cells, finding good agreement for 120 and disagreement for 11, in the process demonstrating specific areas where the biology is not well-understood (Chen et al., 2004).

Sorger applies experimental and computational approaches to the analysis of chromosome segregation, genomic stability and prog-rammed cell death in yeast, mice and human cells. In collaboration with Doug Lauffenburger (MIT) and others, his apoptosis work has included the development of an experimentally-based 400 equation ODE model that can capture cell-type specific variation in the generation of survival signaling emanating from the epidermal growth factor (EGF) receptor (Schoeberl et al., 2003). Bolouri has utilized similar modeling as well as theoretical analysis to study gene regulatory networks, including those important for organism development and the immune system (Ramsey, Orrell, and Bolouri, 2005; Bolouri and Davidson, 2003).

One step up on the continuum of biological scale is cell behavior, frequently modeled in relationship to cell surface receptor dynamics as well as intracellular signaling networks. For example, Jennifer Linderman (University of Michigan) focuses on the dynamics of receptor binding and trafficking and how these influence cell response to endogenous and exogenous ligands (e.g., therapeutic drugs). She has used modeling to demonstrate how receptor desensitization and drug-induced signaling may be decoupled through alteration of drug properties (Woolf and Linderman, 2003). Doug Lauffenburger's (MIT) work has focused on deciphering how cells interpret ligand binding through the dynamics of receptor trafficking and signal transduction to result in a specific behavior, with primary focus on cell proliferation, chemotaxis and apoptosis. A recent study combining modeling of receptor trafficking leading to uptake and degradation of gra nulocyte-colony stimulating factor (G-CSF) by neutrophils with molecular modeling of receptor-ligand structure-function interactions predicted how amino acid substitutions in G-CSF could reduce uptake and thereby increase its half-life within the bloodstream when administered to neutropenic patients, such as cancer patients on chemotherapy (Sarkar, et al., 2002; Sarkar and Lauffenberger, 2003). Hans Othmer (University of Minnesota) has made significant contributions through modeling in a number of biological areas, for example in the chemokinesis and chemotaxis of both single bacteria and populations of bacteria (Albert, Chiu, and Othmer, 2004; Erban and Othmer, 2004). Jason Haugh (North Carolina State University) has used models of the platelet-derived growth factor (PDGF) receptor/PI3-kinase/Akt signaling system in relationship to cell survival to test alternative hypotheses about the dynamic behavior of ligand- receptor interactions. His group demonstrated that dimerization requires the association of two 1:1 ligand-receptor complexes as an initial step with possible formation of stable 1:2 complexes thereafter, rather than dimerization of two receptors by one ligand initially (Park, Schneider, and Haugh, 2003).

The area of metabolic networks and metabolic control within cells is a subset of both of the above but is specialized enough as a field to describe it separately. Because of its special status as an industrially important field, metabolic modeling and analysis was one of the earliest systems biological fields to emerge. The field rests on a foundation of chemical engineering and enzymology that matured in the 1950s and '60s. One of the earliest and most ambitious models was by Garfinkel and Hess in the early '60s and covered hundreds of reactions in the metabolism of Ehrlich ascites tumor cells. Few models of this scale and mechanistic detail have been attempted since. Whole theories have grown up around these and like models to understand the control of flux in these networks. In the early 1970s Heinrich and Rappoport and Kacser and Burns developed Metabolic Control Analysis (MCA), which concerns the sensitivity of steady-state fluxes of metabolic networks to perturbations in enzyme concentrations and other parameters of the system.

Major U.S. contributors in this area include James Liao (University of California-Los Angeles), Gregory Stephanopoulos (Masachusetts Institute of Technology (MIT)) and Bernhard Palsson (University of California-San Diego), among others. Greg Stephanapoulos at MIT has been the leader at using MCA coupled to measurements of metabolites and fluxes to learn how to redistribute material flux in these networks towards desired end products of metabolism. Bernhard Palsson at UCSD has built on classical work in stoichiometric network analysis (SNA) and other flux-based chemical engineering analyses to develop an effective flavor of flux balance analysis that, given fairly con-servative assumptions and a set of input nutrients, can predict the flux through the metabolic network that maximizes growth. He has used these analyses and metabolic reconstructions to predict under which conditions a cell will grow while producing a molecule of interest. He has also used such models and experiments to predict the effect of mutants on the growth mechanism and to derive so-called extreme-pathways that might represent the controllable fluxes in metabolism. Ideker, initially together with Lee Hood, has pioneered data-fusion on molecular interactions, gene expression measurements, and genetic perturbations to build more statistical models of the control of metabolic pathways. Liao's research focuses on mapping control circuitry of metabolism and re-engineering those circuits to provide new functionality within a cell. For example, Liao's group identified the stoichiometric limitation caused by the phosphotransferase system (PTS) in the production of various metabolites, and experimentally demonstrated a solution by overexpression of phosphoenopyruvate synthase to recycle pyruvate back to phosphoenopyruvate (Patnaik and Lao, 1994; Patnaik and Spitzer, 1995).

The cell functions discussed above frequently occur within multicellular organisms or populations of single-celled organisms and therefore mathematical models have been utilized to investigate many biological functions that involve networks of cells, tissues and organs. Denise Kirschner (University of Michigan) has developed ODE and agent-based models of interacting populations of immune cells and infecting microbes to study infectious diseases including HIV/AIDS and tuberculosis. For example, Kirschner's model of *M. tuberculosis* infection in the human lung and draining lymph node was used to predict how different balances of key T cell and macrophage functions would lead to different patient outcomes such as active infection versus latency as well as the biologic functions that could be good targets for modulation by antibiotics (Marino and Kirschner, 2004). Alan Perelson (Los Alamos National Laboratories) has used models to study the dynamics of the T and B cells of the immune system and its response to infection and therapy for infection, for example the treatment of HIV and hepatitis (Gilchrist and Coombs, 2004; Dixit et al., 2004). Building on this model, Leor Weinberger, David Schaffer and Adam Arkin have begun to use such models as the foundation for the design of therapies for control of the onset of AIDS. They proposed an extension to Perelson's model that allowed the exploration of how to best engineer a conditionally-replicating viral gene therapy that would prevent AIDS but not HIV-1 infection. Another area with a long history of modeling is vascular structure and angiogenesis. Thomas Skalak (University of Virginia) has made significant contributions to the understanding of vascular remodeling regulation by mechanical stresses and wound repair using cellular automata models (Pierce, Van Gieson, and Skalak, 2004), while Rakesh Jain (Harvard University) has worked through a succession of ODE and partial differential equation (PDE) models in close conjunction with experimental work to understand the physicochemical drivers of tumor angiogenesis (Stoll et al., 2003; Ramanujan et al., 2000). Mechanical aspects of biological functions at both the macroscopic and microscopic levels are also subjects of modeling. The fluid mechanics of blood flow is a major subject of modeling. For example, Roger Kamm (MIT) is using finite element models of blood flow in the carotid artery in conjunction with magnetic resonance imaging (MRI) and histology to understand how the blood sheer stress correlates with histologic markers in atherosclerotic plaques (Kaazempur-Mofrad et al., 2004).

Electrophysiology of cells and the organization of electrically active cells into tissues and organs is a major subfield spanning molecular networks, cell biology, and multicellular/tissue/organ systems. It has involved so much modeling activity that it deserves its own discussion. Electrophysiology may in fact be the biological area that has the most modeling associated with it, and the studies in this area are on average significantly more quantitative than in most other areas of biology. The electrophysiology of the cardiac myocyte and the organization of myocytes into the structure of the heart are the subject of much research both in the U.S. and internationally, oftentimes through international collaborations. In the U.S., Raimond Winslow (Johns Hopkins University (JHU)), Andrew McCulloch (UCSD), and Yoram Rudy (Washington University), among others, have developed various models of excitation-contraction coupling and its regulation in the cardiac myocyte, as well as integrated models of multiple myocytes into heart tissue, single ventricles, and the whole heart (Winslow et al., 2000; Luo and Rudy, 1994 1; Luo and Rudy 1994 2; McCulloch, Hunter, and Smaill, 1992). Such modeling has been used to better understand how molecular and cellular behavior together with spatial organization determines normal heart function as well as arrhythmias, myocardial infarctio, and other cardiac dysfunctions. Electrophysiology is also central to the function of the nervous system, and modeling is used extensively in this field to understand how various ion channels and pumps drive neural electrical conduction and transmission, as well as how networks of neurons function in organized tissues to result in observable physiology. As the field is quite vast, the reader is referred to several books with review chapters for a broader view of the field (Chow et al., 2005; Koch, 2004; Koch and Segev, 1998; Dayan and Abbott, 2001).

Pattern formation and spatial organization in biological systems involve significant modeling efforts. Many phenomena with important spatial organization aspects occur in organism development. George Oster (UC-Berkeley) has used models to, for example, demonstrate how waves and aggregation patterns in populations of microbes are driven by various characteristics of cell motility (Igoshin et al., 2004). Davidson has developed models to compare alternative hypotheses about the mechanism of the invagination of the sea urchin (Davidson et al., 1995); while Garrett Odell (University of Washington) has studied numerous morphogenesis problems, including how specific genetic control modules drive segment polarity in Drosophila (von Dassow, et al., 2000; von Dassow and Odell, 2002). James Murray (University of Washington) has contributed to the understanding of numerous spatial and patterning phenomena in biology including scarring, fingerprint formation, and skin patterning (Tranquillo and Murray, 1993; Cruywagen, Maini, and Murray, 1994). Like Odell, Stanislav Shvartsman (Princeton University) also combines modeling with gene-tics and cell biology to understand patterning (Pribyl, Muratov, and Shvartsman, 2003 1; Pribyl, Muratov, and Shvartsman, 2003 2).

Finally, it is worth noting that various aspects of whole organism physiology, such as metabolism and respiration have a long history of modeling. These models typically described the biological components and functions at the level of tissues, organs, and/or organ systems including physical properties and geometric features, which fell from favor with the advent of cellular and then molecular biology in the later part of the twentieth century. The work described above in multicellular/tissue/organ networks, which includes more molecular and cellular biology, could be considered the new physiologic modeling when it bridges to aspects of organism function.

Analysis of network organization is most frequently focused on the properties and organization of intracellular biochemical networks based on the large databases arising from genomics, transcriptomics, and proteomics, although the same principles are applicable to networks of cells or other biological elements. This is exemplified by the work of Albert-Laszlo Barabasi (Notre Dame) and colleagues analyzing the properties of many network types, from intracellular proteins to the internet (Yook, Jeong, and Barabasi, 2002; Barabasi and Bonabeau, 2003; Jeong et al., 2000). In studying the connectivity of elements in these networks, he has demonstrated that many such networks are "scale-free." In scale-free networks the probability P of any node being linked to some number k of other nodes follows a power law distribution (log P(k) vs. log k is linear). Slightly different analyses of the properties of these networks lead to predictions of different variations of the scale-free architectures which have implications both for how they are controlled and how they arose evolutionarily. These studies attempt to link various topological properties of the network to properties such as the speed at which information can be communicated, which points in the network are most susceptible to failure, and how different network architectures are more or less robust. The architectural arguments are still somewhat phenomenological (such as noting that highly-connected proteins may have a higher chance of being essential, etc.), while more generic statistical theories come to diametrically opposed interpretations (Carlson and Doyle, 2002; Carlson and Doyle, 1999; Morohashi et al., 2002; Csete and Doyle, 2002; Kitano et al., 2004).

Motifs and modularity are another of the major areas of study of network organization. One of the seminal papers in this area was a mechanistic study of the robustness of exact adaptation in the *E. coli* chemotactic response by Naama Barkai (now at Weizmann Institute) and Stanislas Leibler (Rockefeller University). In this work they demonstrate how the architecture of the signal transduction network ensures exact adaptation of the response regulatory activity to a step of chemoattractant throughout a wide range of kinetic parameters for the underlying biochemical reactions. This robustness was subsequently shown to exist through measurements of *E. coli* with differently expressed chemotactic pathway molecules by Uri Alon and Leibler. Tau MuYi and John Doyle showed that the engineering explanation for this was the existence of an integral feedback motif in the network. The search for "overrepresented" examples of these seemingly important control motifs has become a popular area of study facilitated by better quality databases of cellular networks and high-throughput datasets of molecular interaction. However, caution in the analysis of these motifs from the topological viewpoint is necessary. Elowitz and Leibler have shown experimentally that the same topology of a gene expression motif can yield very different dynamics depending on the exactly kinetic and thermodynamic parameters. This had been predicted theoretically for years but the experimental demonstration was powerful.

The linkage of motifs to evolutionary processes is only just beginning to be explored. The statistical overrepresentation of certain topologies of biochemical interactions is evocative and Chris Voigt, Denise Wolf and Adam Arkin have explored why certain topologies might be selected evolutionarily because of their dynamic flexibility in a study of the *Bacillus subtilis* sinIR operon and made a first attempt to understand the evolutionary selection on different parts of the motif by comparing the sequence of orthologous implementations of the motif in related bacteria. Rao and Arkin examined how small differences in the orthologous chemotaxis pathways in *E. coli* and *B. subtilis*, while having similar gross behavior, differed in the mechanisms of control and the resultant robustness of the network. These approaches to the quantitative analysis of cellular regulatory motifs and the linkage to the evolution of these pathways promised a more complete understanding of the design and architecture of cellular pathways.

Major Alliances, Collaborations and Institutions

It is noteworthy that in the last decade several large alliances and collaborations have been formed that have goals to understand biological systems as integrated systems (e.g., systems biology) and have included mathematical modeling as a central method for the research. Perhaps one of the oldest, a grassroots international effort with many U.S. contributors but without specific federal support, is the Physiome Project (Crampin et al., 2004), spearheaded by James Bassingthwaighte (http://www.physiome.org/). The Physiome Project's major long-range goal is to understand and describe the human organism, its physiology and pathophysiology quantitatively, and to use this understanding to improve human health.

The Alliance for Cell Signaling (AfCS, http://www.signaling-gateway.org), directed by Alfred Gilman (University of Texas Southwestern Medical Center), is a multidisciplinary, multi-institutional research program to study the network properties of cellular signaling systems utilizing a well-organized system for obtaining cell samples and running experiments, databasing results, and integrating the knowledge into models. The AfCS was originally focused on B cells and muscle cells, but recently refocused on macrophages because of significant technical difficulties dealing with the first two. Approximately 50 investigators at 20 academic centers are involved in the AfCS, and it is funded by the National Institute of Health (NIH) and five major pharmaceutical companies. In 2003, the AfCS formed a partnership with Nature Publishing Group to create the Signaling Gateway (http://www.signaling-gateway.org), which provides signaling data and results from AfCS freely to interested parties. Another multi-institution initiative is the Cell Migration Consortium (http://www.cellmigration.org/index.html), which aims to accelerate progress in migration-related research by fostering multi-disciplinary research activities and producing novel reagents and information. The Consortium is comprised of over thirty investigators and collaborators from over 15 institutions, and includes a modeling initiative as one of its key thrusts.

An example of a major center in systems biology at a single institution is the Cell Decision Processes (CDP) Center at MIT (http://csbi.mit.edu/research/projects/celldecision), directed by Peter Sorger and funded by the National Institute of General Medical Sciences (NIGMS) for $16 million over five years. The CDP Center research involves an interdisciplinary team of cell biologists, computer scientists and Microsystems engineers tackling the systems biology of protein networks and signal transduction in mammalian cells, with particular focus on programmed cell death. CDP Center research is based on the hypothesis that understanding cell decision processes requires the development of network models that combine quantitative rigor with molecular detail. The resulting models are hybrids that contain highly specific representations of critical reactions and abstract representations of the system as a whole. Other systems biology centers supported by NIGMS include the Center for Quantitative Biology in Princeton; the Center for Cell Dynamics at the University of Washington; and the Bauer Center for Genomics research at Harvard.

The U.S. government also has a number of departments and centers associated with its research organizations that focus on systems biology and/or modeling of biological systems. These include the systems biology department at Pacific Northwest National Laboratory (http://www. sysbio.org), the physical biosciences division dedicated to quantitative biology at the Lawrence Berkeley National Laboratory, and the systems biology Genome to Life (GTL) projects sponsored by the Department of Energy.

Biological Modeling and Network Organization Analysis in Europe

Biological modeling in Europe has a long history, and is indicative of the wide variety of subjects addressed and techniques being utilized in the field. The specific sites that employ modeling of biology systems in their research that the panel visited in Europe and Japan are listed in Table 4.1. Here a number of studies that were notable for their innovation and contributions in the different categories listed above are described.

A particularly notable study of a biochemical network is the work of Ursula Klingmüller of the German Cancer Research Center in Heidelberg and a leader within the German Hepatocyte Project. She and her colleagues' work on the Jak-stat regulatory network has lead to new insights about its structure and function. She utilized a combined modeling-experiment approach to distinguish between two alternative hypotheses about how signal transducer and activator of transcription (STAT), after phosphorylation by a Jak-activated receptor, could enter the nucleus and activate gene expression (Swameye et al., 2003). They used a novel procedure to compare mass-action kinetic models of the alternative hypotheses to experimental data to demonstrate that STAT5 must become unphosphorylated and recycled out of the nucleus and back in again rather than get trapped in the nucleus and be degraded. They then used the model to ask what intervention is most effective for increasing STAT5-p in the nucleus, the conventional view of increasing phosphorylation, or blocking nuclear export. The model predicted the latter, and they used experiments to demonstrate this was correct.

Intracellular metabolic networks and control are areas with significant attention in Europe. David Fell at Oxford Brookes University in England utilizes metabolic engineering approaches to study metabolism primarily in plants and bacteria, including threonine biosynthesis in *E. coli*, potato tubor metabolism (in collaboration with Advanced Cell Technologies in Cambridge), photosynthesis, and antibiotic production in actinomycete (Poolman, Assmus, and Fell, 2004; Schafer et al., 2004; Chassagnole et al., 2003; Thomas et al., 1997). In addition, he utilizes structural modeling, the deconstruction of large networks into smaller substructures, to predict pathways that are feasible from gene expression information and those capable of greatest metabolic yields. In one notable study, he predicted that there should be six categories of arid-environment plants in terms of crassulecian acid metabolism instead of only four as previously known. Recent recognition of a fifth and discovery of a sixth lends support to his prediction.

Hans Westerhoff of Free University in Amsterdam, Netherlands, is a leader within a large group of researchers in the Netherlands that utilize metabolic control analysis to study metabolic networks and regulation in yeast and bacteria as well as signal transduction. In related work, the focus of Reinhart Heinrich of Humboldt University, Berlin, is on dynamic models of metabolism and control of networks as well as other biological pathways. His approach involves modeling, with close verification by experiments, using methods from nonlinear dynamics and simulation, bifurcation theory, metabolic control analysis, stoichiometric network analysis, stochastic process theory, optimization and graph theory. In recent work with Mark Kirschner combining modeling and experimentation, he studied the effect of Wnt stimulation on β-catenin expression in *Xenopus* oocyte extracts, finding that depending on topology there were greater and lesser regions of stability of the G-protein signal transduction network based on both the number of kinases in the network and phosphatase activity (Lee et al., 2003). In other work, Heinrich has analyzed the evolution of networks by using the large metabolic maps in the Kyoto Encyclopedia of Genes and Genomes (KEGG) database (http://www.genome.jp/kegg), determining how such metabolism could be built a step at a time from finding critical routes from each metabolite to other necessary endpoints. Different substrates can make different numbers of primary metabolites through the known reaction network. For example, adenosine triphosphate (ATP) can make about 1,500 whereas glucose can only make around 50, suggesting that a great deal of the cell's metabolic network could have been elaborated from a smaller metabolism based only on transformations of ATP (Ebenhoh, Handorf, and Heirich, 2004).

Cell function and its intracellular regulation are the subjects of interest of a longstanding modeler in Europe, Albert Goldbeter at the University Libre de Bruxelles in Belgium, a university with a long tradition of theoretical biology. Through his modeling work Goldbeter has contributed to the understanding of various dynamic cellular phenomena including regulation of Circadian rhythms and metabolic oscillations (Goldbeter et al., 2001; Leloup and Goldbeter, 2003; Goldbeter, 2002). Others in his department have contributed in the areas of dynamics of regulatory gene networks, calcium signaling, theoretical ecology and social insect behavior.

Cellular chemotaxis of both prokaryotes and eukaryotes is another subject with a significant history of modeling and analysis to help understand the molecular and physical mechanisms driving it. Dennis Bray, Cambridge University, has made significant contributions on chemotaxis of *E. coli* since the early 1990s. This is a particularly tractable system to study because only six or seven proteins are involved in its regulation. Bray's group has both deterministic and stochastic models they use to study excitation, adaptation, mutant phenotype, individuality, and the effect of architecture of intracellular space (Lipkow, Andrews, and Bray, 2005; Shimizu, Aksenov, and Bray, 2003; Bray and Bourret, 1995). Another European researcher modeling chemotaxis and cell motility is Wolfgang Alt (Bonn, Germany), although the panel did not visit his laboratory.

In the subfield of electrophysiology, Denis Noble of Oxford University has studied cardiac biology for more than forty years, collaborating with Raimond Winslow (U.S.), Peter Hunter (New Zealand), and Andrew McCullough (U.S.). Noble's research group tightly integrates experimental and modeling approaches in their study of how ionic currents drive cardiac myocyte function and how cellular function is integrated spatially and dynamically to drive the function of the whole heart (ten Tusscher et al., 2004; Garny et al., 2003; Markhasin et al., 2003; Noble, 2002). Over the years, his models have contributed to the understanding of energy conservation, the necessary stoichiometry of ion exchangers, and mechanisms of calcium balance in cardiac myocytes, and the implications of these for cardiac dysfunctions such as arrhythmia (see Noble, 2002, for review). The heart is arguably the organ most comprehensively modeled, in terms of biological detail (molecular, cellular, spatial organization, dynamic function). Noble notes that what has made this possible is that relevant experimental work has been ongoing for 40 years, providing a vast body of data and knowledge. The major regulators of the cell and tissue function (ion channels generally) are quite accessible to measurement, and the cell properties that contribute to many aspects of whole organ function are relatively few and aren't strongly dependent on vast intracellular signaling networks.

Ernst Dieter Gilles and his associates at the Max Planck Institute for Dynamics of Complex Technical Systems in Magdeburg, Germany have focused on aspects of model validation and model-based experimental design (Kremling, 2004), areas with a long history in engineering and physical sciences but less in biological modeling. Gilles' department closely integrates experimental and modeling work for the purpose of improving the understanding of biological phenomena and identifying solutions for medical problems, particular drug target identification. The methods employed included detailed mathematical modeling, model validation and iterations (via design of experiment) for hypothesis testing, system-theoretic analysis of properties including robustness, and decomposition and model reduction. The particular biological areas of research include signal transduction and regulation in bacterial cells and eukaryotes, metabolic network structure, and computer-aided modeling and analysis of cellular systems (Stelling and Gilles, 2004; Schmid et al., 2004; Stelling et al., 2004). Gilles has noted that their initial attempts at collaborations with biologists were not effective because their engineering approaches were not appreciated by the biologists. They have become more effective over the years by refining their modeling efforts, becoming more visual and including lower-level biological details. Their work has made important contributions in the extraction of design principles from the robustness analysis of the circadian clock (Stelling, Gilles, and Doyle, 2004), and the use of identifiability tools to guide experimental design for modeling of cell cycles. An important conclusion of the latter work was that perturbations were more important than additional measurements in formulating an optimal experiment for model identification.

Large-Scale Alliances and Collaborations

Several large-scale systems biology projects in Europe include modeling as a cornerstone. One launched in January 2004 is the Hepatocyte Project in Germany, funded by the German Federal Ministry of Education and Research (BMBF) in its "Systems for Life—Systems Biology" initiative. The thematic focus and structure for the Hepatocyte Project was shaped starting in 2001 by an expert panel of about 80 scientists through four workshops. The goal to understand the mechanisms of behavior of hepatocytes was ultimately selected because of the hepatocyte's central function in metabolism, their central role in the uptake and conversion of drugs and thereby their interest to industry, and their ability to regenerate. The panel expected this research to have high impact on problems in pharmacology and pathophysiology, despite the challenges involved in the complex biology of hepatocytes, the difficulty in their handling and cultivation, and need to create a bioinformatic and modeling framework to organize information about them. The Hepatocyte Project has an interdis-ciplinary competence network linking bioscience with computer science, mathematics and engineering sciences. Two sub-projects (Project A is detoxification and dedifferentiation in hepatocytes and Project B is regeneration of hepatocytes) rest on two technological platforms, cell biology and modeling. The project has 25 participating groups, and funding of €14 million is provided over three years beginning in 2004.

Another large consortium project entitled "Biosimulation: A New Tool in Drug Development" (http://chaos.fys.dtu.dk/biosim/ Beskrivelse_af_BioSim.html) was announced in December, 2004 by The European Commission and the Technical University of Denmark (DTU) to support the growing importance of modeling for biomedical research, and pharmaceutical development in particular. The project is funded under the European Union's Sixth Framework Program for Research at the level of €10.7 million over five years. The project's aim is to strengthen Europe's competitiveness within drug development by bringing together the leading European biosimulation experts in a scientific network and promoting collaboration across disciplinary boundaries as well as between industry, regulatory authorities, and academia. The network will focus on the development of professional, physiologically based models that can help the pharmaceutical industry develop safe and effective drugs at significantly lower costs. It was motivated by the recognition that academic institutions in Europe have significant expertise in biological modeling, and several groups are individually at the research front in their specific areas, but the research is strongly fragmented and the industry itself has relatively few qualified experts in the field. Coordinated by the DTU, the network comprises approximately 100 researchers from 25 universities/research centres, nine small or medium-sized enterprises, the medicines agencies of Denmark, Spain, the Netherlands and Sweden, and one large pharmaceutical company, Novo Nordisk.

Biological Modeling and Network Organization Analysis in Japan

Modeling of biological systems is quite evident in Japan, although from the sites the panel visited it appears to be less extensively utilized than in the U.S. and Europe. The sites with modeling components of their research that the panel visited are included in Table 4.1 and a few of those are highlighted here.

The work of Satoru Miyano focuses on estimating gene networks from genome-wide biological data, as well as software tools for bioinformatics and modeling (described below) and pathway database projects. Miyano is the current president of the Japanese Society for Bioinformatics, and the editor-in-chief of the newly established IEEE Transactions on Computational Biology and Bioinformatics. This group has developed hybrid functional Petri net methods for gene network inference (Akutsu, Miyano, and Kuhara, 2000; Doi et al., 2004) and their current Gene Network Inference Method (G.NET) can yield the optimal gene network for twenty genes (on a Sun Fire 15K 100 CPU machine) in one day. As a test case they used the method to discover the function of the oral antifungal griseofulvin and predict new targets related to it. Specifically, they measured gene expression as a function of exposure, and used a Boolean network approach using the drug as a virtual gene to predict genes that were directly affected by the drug. Then they made predictions of related components in the pathway that would also be effective if manipulated (Savoie et al., 2003). This work was done in collaboration with a small company with compounds related to griseofulvin.

A well-known modeling group in Japan is that of Hiraoki Kitano of the Symbiotic Systems Project associated with the Systems Biology Institute (SBI). The basic goal of the project is to develop and apply new technology and computational tools to understand dynamical phenomenon in cellular systems. Areas of research include the theory of the robustness of cellular networks, signal transduction in yeast and mammals (collaborating with the Alpha Project and the Alliance for Cellular Signaling), respiratory oscillations in yeast, and calcium oscillations in mammalian cells (including collaborations with the Karolinska Institute) (Ktano et al., 2004; Kitano, 2004; Yi, Kitano and Simon, 2004). A significant focus of Kitano's work has been the development of modeling software and markup languages to encode and share models as described below, as well as advanced hardware platforms for simulations.

Table 4.1
European and Japanese Sites Utilizing Modeling Visited by Panel

Institution	Location	Principal Investigator(s)	Research Subjects
Department of Physiology, Oxford University	Oxford, England	Denis Noble, Peter Kohl, Ming Lei	Cardiac biology; respiratory biology
Department of Anatomy, Cambridge University	Cambridge, England	Dennis Bray	Chemotaxis in *E. coli*
University of Sheffield (multiple departments)	Sheffield, England	Chris Cannings, Richard Clayton, Steve Dower, Mike Holcombe, Nick Monk, Eva Qwarnstrom, Francis Ratnieks, Rod Smallwood, Phillip Wright, Will Zimmerman	Cell organization in tissues; system organization in insect societies; protain-protein interaction networks; inflammatory mediator signaling; agent-based models; toll receptor signaling; cell signaling and network pattern formation; ventricular fibrillation
Mathematics Institute, University of Warwick	Coventry, England	Andrew Millar, Nigel Burroughs, Jim Beynon, Greg Challis	Network regulation; transmembrane protein transport; Circadian clock; immunology; population genetics
Centre for Mathematical Biology/Mathematical Institute, Oxford University	Oxford, England	Philip Maini (Director), Jon Chapman, Chris Scofield, Jotun Hein	Nutrient and drug delivery to tissue with application to cancer; cardiovascular biology; cell cycle; model integration methods; population genetics and genomics; wound healing

Table 4.1
European and Japanese Sites Utilizing Modeling Visited by Panel

Institution	Location	Principal Investigator(s)	Research Subjects
School of Biological and Molecular Sciences, Oxford Brookes University	Oxford, England	David Fell	Metabolic networks; cell cycle; IκB regulation
Centre for Mathematics in the Life Sciences and Experimental Biology, University College London	London, England	Anne Warner (Director), Anthony Finkelstein, Jonathan Ashmore, Robert Seymour	Liver metabolism; software and methods to integrate across biological scales
Computational Systems Neurobiology program, European Bioinformatics Institute	Hinxton, England	Nicolas LeNovere	Topology and dynamics of neuronal cell signaling pathways; dopamine signaling
German Cancer Research Center/Hepatocyte Project	Heidelberg, Germany	Otmar Wiestler, Siegfried Neumann, Ursula Klingmüller, Willi Jager, Wolfgang Driever, Matthias Reuss, Eric Karsenti, Jens Doutheil, Jan Hengsteler, Sven Sahle	Hepatocyte Project; signaling pathways; structural and functional genomics; cancer risk factors and prevention; tumor immunology; innovative diagnostics and therapy
Max Planck Institute for Dynamics of Complex Technical Systems	Magdeburg, Germany	Ernst Dieter Gilles, Jorg Stelling	Model validation and experiment design; signal transduction and regulation; computer-aided modeling and analysis of cellular systems
Collaborative Research Center for Theoretical Biology, Humboldt University	Berlin, Germany	Reinhart Heinrich, Hanspeter Herzel, Peter Hammerstein, Hermann-Georg Holzhutter	Metabolic control; biological dynamics; Ras signaling; Circadian clock; Huntington's disease; Hepatocyte Project
Department of Vertebrate Genomics, Max Planck Institute for Molecular Genetics	Berlin, Germany	Hans Lehrach, Edda Klipp, Silke Sperling	Yeast stress response and mitochondrial damage; Downs syndrome; cardiac development;
University Libre de Bruxelles	Brussels, Belgium	Albert Goldbeter	Biological dynamics, Circadian rhythms, cell cycle
Vrije Universitaet (Free University)	Amsterdam, Netherlands	Hans Westerhoff, Jurgen Haanstra, Frank Bruggemann, Jorrit Hornberg	Metabolic control analysis, network based drug design, Silicon Cell toolkit, signal transduction

Table 4.1
European and Japanese Sites Utilizing Modeling Visited by Panel

Institution	Location	Principal Investigator(s)	Research Subjects
Delft University	Delft, Netherlands	Wouter van Winden	Metabolic control
Cell/Biodynamics Simulation Project, Kyoto University	Kyoto, Japan	Akinori Noma (Director), Tetsuya Matsuda, Nobuaki Sarai	Cardiac biology; biosimulation software development
Symbiotic Systems Project, Systems Biology Institute	Tokyo, Japan	Hiraoki Kitano	Modeling technology; model standards technology; robustness of cellular networks; yeast signaling
Human Genome Center, Institute of Medical Science, University of Tokyo	Tokyo, Japan	Satoru Miyano	Gene network inference from genome-wide data; yeast networks; pathway databases; software for bioinformatics and simulation
Department of Computational Biology, Graduate School of Frontier Sciences, University of Tokyo	Tokyo, Japan	Shinichi Morishita, Takashi Ito	Functional genomics and signaling in budding yeast; mammal epigenomics; computational approaches to Omics
RIKEN Yokohama Institute	Yokohama, Japan	Akihiki Konogaya, Mariko Hatakeyama, Shuji Kotani	ErbB-mediated signal transduction; yeast cell cycle; reaction-diffusion systems; Grid computing; cell cycle modeling
Institute for Advanced Biosciences, Keio University Tsuruoka Campus	Yamagata, Japan	Masaru Tomita, Hirotada Mori	E-CELL project
Bioinformatics Center, Institute for Chemical Research, Kyoto University	Kyoto, Japan	Minoru Kanehisa	KEGG project; dynamic metabolic models

Modeling of electrophysiological phenomena, with flavors of intracellular signaling, cell behaviors, and multicellular/tissue/organ categories, is the work of Akinoi Noma at Kyoto University. Noma has had a distinguished career in experimental cardiac physiology and electrophysiology and only added modeling to his research methods in the past several years. His Cell/Biodynamics Simulation Project is focused on developing and applying models of cardiac myocytes and their integrated function in the heart, closely related to work by Noble described above. Their novel contributions include the addition of intracellular biochemical mechanisms such as ATP utilization, redox state balance, and pH (Matsuoka et al., 2004). In contract, most myocyte models have concentrated on cell surface molecular entities and behaviors, specifically the ion channels, pumps and action potentials. This is a National Leading Project for Cooperation between Industry and Academia sponsored by Ministry of Education, Culture, Sports, Science and Technology (MEXT) on a five-year grant, which requires industry involvement and work that is relevant to economic growth. Dr. Noma has involved seven pharmaceutical companies to date, including Nippon Shinyaku, Shionogi, Sumitomo Chemicals, Tanabe, Sankyo, Takeda, and Mitsubishi Well Pharma, as well as researchers at Kyoto and Keio Universities in Japan and others in Poland and Korea. Four of the collaborating companies have placed a full-time employee in the Dr. Noma's group in Kyoto, while the others send visiting researchers for short visits periodically. This project is also a major developer of modeling/simulation tools, as described in infrastructure below.

In the area of network organization analysis, the work of Masanori Arita (University of Tokyo) has demonstrated how structural infor-mation of metabolites is important for computing biochemical pathways and understanding the network properties of those pathways (Arita, 2004).

INTEREST AND INVOLVEMENT OF INDUSTRY

The interest and involvement of industry in systems biology efforts that include modeling and network organization work are significant although hardly ubiquitous, and quite variable between regions. The industry of concentration is pharmaceutical and biotechnology, although applications in the nutraceutical, agricultural supply/chemical and bioprocess industries also exist. Here activities within the industry as well as the relationship between industry and academia are summarized.

R&D Using Modeling in Industry

Companies using mathematical models of biological systems are treating them essentially as *in silico* laboratories that complement the experimental laboratories. As described in the introduction, models are particularly suited to tracking the states of and relationships among numerous elements through time and space, especially for large, complex systems as found in biology and medicine. Applications of systems biology modeling within the pharmaceutical industry vary from drug target identification to clinical trial design and analysis. Models of intracellular biochemical networks are typically utilized to investigate how modulation of that network, by agonizing or anta-gonizing network components, affect the associated cell function. The direct relationship of that cell function to a particular disease is disease-specific, for instance proliferation of a cancer cell is directly related to disease outcome (tumor size, say), whereas modulation of an inflammatory cell within the asthmatic airways still requires extrapolation to relevant clinical endpoints. Models of whole organs in which a disease is isolated, like the heart for cardiac arrest or arrhythmia, or of multicellular/tissue/organ systems that include relevant clinical endpoints, bridge cellular function to relevant clinical outcomes much as an experimental animal model would.

Given that drug targets are nearly always molecular, modeling within the pharmaceutical industry is most frequently focused at the level of intracellular biochemical networks. Several companies in the U.S. that are focused on development and use of such intracellular network models include Gene Network Sciences (U.S.), Merrimack Pharma-ceuticals (U.S.), Genomatica (U.S.) and Physiomics (U.K.). Gene Net-work Sciences uses inference modeling and mechanistic simulation of intracellular biochemical and gene networks related to cell cycle to do research on cancer and its treatment (Christopher et al., 2004). More recent directions include cardiac electrophysiology. Physiomics also focuses on modeling cell cycle control. EGF receptor dynamics and downstream signaling associated with various cancers are the subject of modeling at Merrimack Pharmaceuticals. In contrast, Genomatica utilizes models of microbial and yeast cell metabolism to improve bioproduction of chemicals and proteins, among other applications.

Multicellular/tissue/organ network function has also attracted the application to pharmaceutical development. Entelos, Inc. (U.S.) utilizes dynamic ODE models of the biological systems involved in specific diseases to evaluate drug targets, select lead compounds, predict biomarkers, and design clinical trials. Current areas of work include model representations of human diseases (asthma, hematopoiesis, obesity, rheumatoid arthritis, and type 2 diabetes), and animal models of human disease (type 1 diabetes).

Larger pharmaceutical companies that utilize biological systems modeling in some of their R&D activities include Pfizer, AstraZeneca, Hoffman-La Roche, Johnson & Johnson Pharmaceutical Research Division, GlaxoSmithKline, Novartis, Organon, and Bayer. This list is probably not exhaustive. In some of these companies the inclusion of modeling is quite extensive, while in many it is often confined within a single therapeutic area or a single group that works with multiple experimentalists.

There seem to be fewer companies primarily devoted to using modeling as a primary R&D method in other areas of the world. In addition to Physiomics mentioned above, another modeling-focused company is

Optimata in Israel, whch utilizes modeling to optimize drug dosing and schedules, in particular for cancer treatment.

While companies are notably less forthcoming than academic researchers with making their models and research results public, there is emerging evidence that modeling is significantly impacting pharmaceutical development. For example, Entelos and Organon recently made public that they are engaging in collaborative drug development focused on three novel targets that were identified using Entelos Rheumatoid Arthritis PhysioLab® platform (http://www.entelos. com/news/ pressArchive/press62. html). Johnson & Johnson Pharmaceutical Research and Development has also disclosed that simulations of a type 2 diabetes drug with Entelos Metabolism PhysioLab platform enabled them to reduce the patient recruitment requirements by 60% and trial duration by 40%, as compared to the originally proposed trial protocol (Trimmer et al., 2005). Optimata has also made public that they are utilizing their simulation methods to create individualized treatment protocols for breast cancer patients in a clinical trial at the Nottingham City Hospital Trust, although results are not yet available for the study (http://www.export.gov.il/Eng/_Articles/Article.asp?CategoryID=464&ArticleID=1017). An example of results from a large pharmaceutical company is Hoffman-La Roche's use of modeling of a treatment for hepatitis C. They used modeling and simulation to account for a variety of factors in different patient populations such as genotype of virus and weight of the patient. The results were important for the approval of the drug in both Europe and the U.S. (McGee, 2005).

Relationships between Academia and Industry in Different Regions

Close relationships between industry and academia were particularly obvious in Japan in comparison to the U.S. and Europe. Many Japanese academic researchers stated that the government research funding agencies strongly encouraged collaborations with industry and the transfer of technologies to industrial concerns, either through the start up of new companies or to established companies. In numerous laboratories there was active involvement in the research by industry staff in residence.

In contrast, in Europe the relationship of industry to academic research was at the level of a few collaborations (without exchanging personnel) and the encouragement by funding agencies to transfer findings and technology developments to industry. The recent large collaborative projects in Germany and the EU described above were both influenced by the interests of the pharmaceutical industry, although in the Hepatocyte Project there is no actual industry involvement. The EU Biosimulation project is too new to know how involved the corporate partners will become beyond providing funding.

In the United States, academic-industry collaborations are common, although it is uncommon for industry personnel to spend time in an academic lab. The reverse is probably more likely. It has become the norm for universities in the U.S. to patent research findings and then license them to companies or for the inventors to start new companies to commercialize the inventions.

INFRASTRUCTURE SUPPORTING MODELING AND NETWORK ORGANIZATION ANALYSIS

Throughout the U.S., Europe and Japan, the study panel found significant efforts devoted to the development of software platforms in which to build and simulate mathematical models of biology (and sometimes more general) systems. While variations exist among these platforms, the panel concluded that significant replication of numerous features and capabilities among them also exist. A table listing a number of modeling and simulation platforms focused on or frequently used for modeling of biological system networks is given in Table 4.2 along with some descriptive information and web sites at which the reader can learn more. Additional lists on the web can be found on the Systems Biology Markup Language (SBML) web site (http://sbml.org) for software packages that are compatible with SBML and the Bio-SPICE web site (https://users.biospice.org/tools.php) for those that are part of the Bio-SPICE project.

Table 4.2
Software for Modeling and Simulation of Biological Systems

Software Platform	Main Applications	Developer or Main Contact	Availability to Others[1]	Web site
Bio-SPICE	Collection of many software packages with many applications—see web site	DARPA-sponsored consortium; Sri Kumar, program manager	Source and binary download via web site	https://users.biospice.org/home.php
Virtual Cell	ODEs with multiple compartments; PDEs	Leslie Leow, U.S.	Use through internet	http://www.vcell.org
Teranode Design Suite		Teranode Corp.	Commercial; reduced price for academic use	http://www.teranode.com
MATLAB & SimuLink	General math simulation tool	The MathWorks	Commercial; reduced price for academic use	http://www.mathworks.com

Table 4.2
Software for Modeling and Simulation of Biological Systems

Software Platform	Main Applications	Developer or Main Contact	Availability to Others[1]	Web site
Mathematica	General math equation solver and simulation	Wolfram Research, Inc.	Commercial; reduced price for academic use	http://www.wolfram.com
YAGNS (Yet Another Gene Network Simulator)	Biochemical reaction network simulator (ODEs)	RIKEN Yokohama, Japan	Access via web upon request	http://big.gsc.riken.jp/big/Research/Cellular_Knowledge_Modeling_Team/Folder.2004-01-15.5608/Folder.2004-01-15.5713/Document.2004-01-15.3211
Genomic Object Net / Cell Illustrator	Biological pathway modeling and simulation based on hybrid functional Petri net (HFPN) and XML	Gene Networks Inc., Japan	Contact company	http://www.genomicobject.net/member3
Cell Designer	Structured diagram editor for drawing gene-regulatory and biochemical networks; simulation by linking to other packages	Hiraoki Kitano, Japan	Download via web site	http://www.celldesigner.org
SimBio / DynaBioS®	Cell electrophysiology and Finite element modeling of electorphysiologic tissue (heart)	Akinori Noma, Japan	Download via web site	http://www.sim-bio.org
JDesigner/Jarnac	Biochemical network layout tool and simulation package (ODEs)	Systems Biology Workbench project, Japan and U.S.	Download via web site	http://www.sys-bio.org
ProMoT/DIVA	Object oriented and equations based modeling tool for simulation; differential and algebraic equations	Martin Ginkel, Germany	Download via web site	http://www.mpi-magdeburg.mpg.de/de/research/projects/1002/comp_bio/promot
GENESIS/ Kinetikit	Graphical simulation environment focused on signaling networks	Sharat Vayttaden and Upinder Bhalla, India	Download via web site	http://stke.sciencemag.org/cgi/content/full/sigtrans;2004/219/pl4/DC1

Table 4.2
Software for Modeling and Simulation of Biological Systems

Software Platform	Main Applications	Developer or Main Contact	Availability to Others[1]	Web site
Copasi	Complex pathway simulator (not biology specific)	Pedro Mendes, U.S. and Ursula Kummer, Germany	Download via web site	http://www.copasi.org
Cellerator (a Mathematica package)	Mathematica package for automatic equation generation and simulation for signaling networks and networks of cells	Bruce Shapiro, Eric Mjolsness, U.S.	Download via web site	http://www.cellerator.info
BioNetGen	Cell signaling networks based on interactions of individual molecules	Michael Blinov, James Faeder, William Hlavacek, U.S.	Download via web site	http://cellsignaling.lanl.gov/bionetgen
E-Cell	Object-oriented software suite for modeling, simulation, and analysis of large scale complex systems	Masaru Tomita, Japan	Download via web site	http://www.e-cell.org
JigCell	Modeling of biochemical reaction pathways	Virginia Tech	Download via web site	http://jigcell.biol.vt.edu
MCell	Monte Carlo simulator of cellular microphysiology	Thomas Bartol, Jr., and Joel Stiles, U.S.	Download via web site	http://www.mcell.psc.edu
COR (Cellular Open Resource)	Cell electrophysiology	Denis Noble, U.K.	To be available via web site	http://cor.physiol.ox.ac.uk

1 Terms of availability frequently differ depending on the expected use of the software (e.g., non-commercial or commercial) and may require licenses for other software used in the package.

Reasons given by researchers for developing new platforms included the need to have faster simulation capabilities, improved usability, and features specific to the biologic system being modeled by the developer's research group. Usability issues were mentioned with particular attention to making models and simulation more accessible to non-expert users, although technical usage was also mentioned. The latter includes such things as ease of specifying model equations (e.g., specifying equations by drawing structured diagrams rather than typing them), input and modification of parameter values, specification of protocols to simulate, and storage/retrieval of simulation specifications and results. Features desired were typically specific mathematical methods such as those to handle stochastic processes or finite element algorithms, as well as analytical methods such as parameter optimization or sensitivity analysis.

Some modeling software is available commercially, the most commonly used (at least in the U.S.) general purpose numerical computing platform being MatLab by The MathWorks, followed perhaps by Mathematica (Wolfram Research). Matlab and Mathematica both have announced special tools for systems biology. More specialized commercial software includes Berkeley Madonna, and a number of pharmacokinetic simulation packages. Benefits of such commercial platforms are their formal quality assurance/quality control (QA/QC) processes as well as formal means for reporting bugs and requesting new features. The disadvantage is that they are not free, although the cost of most commercial packages tends to be nominal for academic use. Many of the mentioned packages are not specialized for biological systems, are not well linked to biological databases and data sources, and are not user-friendly for the biological user. For the many platforms being developed noncommercially within research groups, the most common method of dissemination (when available) is via the group's web site. The advantage of development within research groups is that the software features and user interface can be closely guided by end-users within the group. Some disadvantages, however, are that QA/QC procedures for such software are frequently unclear and likely absent in many cases and "services" to users outside the developer's group, such as ways to report bugs and request new features, are not always clear and response by developers not assured.

The fact that so many new packages are in development strongly suggests that those available, commercial or otherwise, are not fully meeting significant needs of the biological modeling community. While no one piece of software will meet the needs of all modeling efforts, the panel believes that the community would be well-served by a national or international resource devoted to making broadly applicable platforms widely and freely available, as well as supporting maintenance and expansion. Such a resource should reduce the ongoing proliferation of somewhat duplicative and quite expensive software development. The Bio-SPICE program funded by Defense Advanced Research Projects Agency (DARPA) for the last three years is an example of such a program although it focused primarily on initial development for an open-source infrastructure for integrating such software. The funding for this project ends in 2005 and currently there are no allocated resources from DARPA or elsewhere to continue funding to support dissemination of the resulting software or its continued development.

Sharing of Models

Historically models have been shared through literature publication. As modeling has become a more common way to do biologic research and the models themselves have become larger, this method has become less satisfactory to many and a strong desire to more easily share models in electronic form between research groups has grown. While some researchers simply want to use or modify published models without having to regenerate computer code to do so, others also want to integrate others' models with their own to create models of larger biological systems.

The main difficulty in sharing models electronically is that models are encoded for specific software and hardware platforms that aren't universally compatible with software or hardware in other labs. To alleviate this problem, two main international efforts are underway to better enable model sharing, namely, the development of markup languages that encode mathematical equations typically used in models of biological systems. For encoding models, Systems Biology Markup Language (SBML; http://sbml.org) (Hucka et al., 2003) and CellML (Cell Markup Language; www.cellml.org) are two major languages under development. The idea is to create a language analogous to HTML (hypertext markup language), the common encoding language for the web. As long as one has an HTML "decoder" on one's computer, e.g., a browser such as Microsoft Internet Explorer, then one can interpret the text, pictures, etc. encoded in an HTML file and the same file can be viewed on all computers. SBML is focused on language to encode math that describes biochemical reaction networks while CellML is focused on describing cellular components and compartments as well as biochemical reaction networks. SBML began in Japan and at CalTech and has since become an international collaboration effort with funding from the U.S., Japan and the U.K. Development efforts on CellML are based in New Zealand. At least one other biological modeling-focused markup language is being developed by Satoru Miyano's group in Japan, labeled Cell System Markup Language, although it does not appear to be publicly available yet.

Part of the effort associated with both SBML and CellML is the development of a public repository of models online in the markup language (see http://sbml.org/models.html and http://www.cellml.org/ examples/repository/index.html, respectively). Authors of models are encouraged to deposit a version of their model on the web site, available for download by others. Another such repository just launched in April, 2005, is Biomodels.net (http://www.biomodels.net/), which is supported by multiple organizations from several countries. Numerous individual labs also provide either models for download or for simulation over the Internet via their web sites.

While none of these markup languages are as yet accepted as the standard, researchers (at least in the U.S.) generally acknowledge SBML as being the most frequently used. Certain U.S. funding agen-cies, including the National Science Foundation (NSF), have taken the position that models developed with their funding must be made publicly available in SBML. The panel is not aware of such a policy at other agencies or nonprofit funding groups in the U.S. or by funding agencies in other countries. Journals generally do not yet require that authors provide electronic versions of their models to the readership in any form.

NEEDS AND RECOMMENDATIONS

In this study the panel has recognized several needs and deficits of the modeling efforts in systems biology and makes several recommendations to alleviate these. First and foremost, the panel notes the need for substantially greater and more widespread integration of modeling and experimental programs. Currently the majority of modeling and experimental efforts on a given subject are performed som-ewhat or completely remotely from each other. This decreases the benefit of both to each other, and slows progress in developing new understanding in many fields. Data from diverse laboratories and with diverse protocols are often found to be difficult to compare when placed in model context implying both that models can help ensure consistency among datasets thus preventing spurious conclusions about the significance of a particular observation and that data quality control is even more important than model quality control. While it may seem from the highlighted examples above that modeling work is commonly integrated with experimentation, the examples were selected in part because they demonstrated how well-integrated programs provided unique insights into the subjects of study. The reason that such tight integration is promoted by the panel is because models can't be developed or tested without experimental data, and the experimentation that provides the necessary data is often not obvious without the model guiding its design. The experiment-model iteration paradigm that is most productive is illustrated in Figure 4.3.

Closely related to the need for experimental-modeling integration within research endeavors is a need for improved means of comparing experimental data and modeling results. Many software platforms for modeling currently don't support easy representation of experimental data, for instance, so both the data and modeling results have to be exported to a third software to allow the comparison.

Another major need is for a means of disseminating and maintaining good cheap or free software appropriate to modeling of various problems nationwide and internationally. The proliferation of modeling software platforms was discussed above. It is the panel's conclusion that many of these platforms are duplicative and the time and cost being expended to develop all of them could be better used to make some of them more widely available as well as to maintain and expand those platforms.

The panel also agrees with the widely stated desire of researchers to be able to more easily share their models with one another in electronic form. The panel therefore supports the efforts of the groups developing SBML and CellML, however, these are efforts funded by several grant agencies, that funding is not guaranteed, and no one is required to use either of these (or other translational languages), so their future, and the possible future of improved ease of sharing models, is not assured.

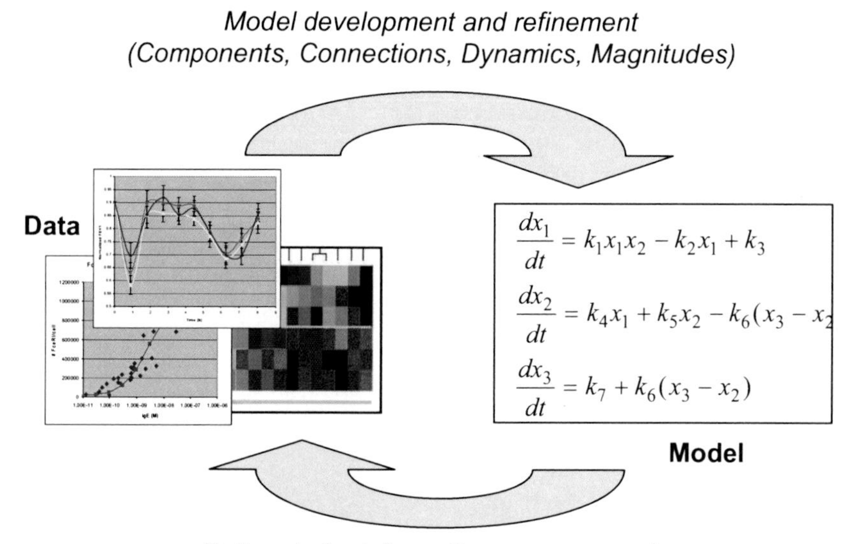

Figure 4.3. With the purpose of increasing understanding of a biological system function, one needs a set of data to develop the first model. That first model can then be simulated or analyzed to pinpoint uncertainties that are important to the system and then recommend a new set of experiments to measure relevant quantities to reduce those uncertainties. The new data can be used to revise the model. The model at various stages of iteration can also be used to test hypotheses about the system's function or means of modulating that function, and interesting predictions from those tests can then also be verified (or refuted) experimentally.

SUMMARY OF KEY FINDINGS

This chapter described the state of systems biology research involving modeling and network organization analysis. A number of key findings can be summarized. Modeling and network organization analysis efforts are utilized in many areas of biological study and in all countries visited, but are definitely not ubiquitous throughout biological and biomedical research. The panel found that research efforts that closely integrated modeling with experimental work were the most productive in terms of driving new understanding of a biological system. Related to this, the panel concluded that a substantial increase in the number of efforts using model-based experimental design is needed to attain the most informative data, which leads to maximally useful models. An implication of this is that having large data generation centers to globally profile molecular abundances or activities might not provide the ideal substrate for gaining a mechanistic/causal understanding of how cells transform genotype into phenotype. Data generation and models that integrate, follow impli-cations of, and make testable assertions about the causal basis of that data need to be strongly linked. In addition, better tools for model-experiment comparison would be helpful. Significant resources are being invested in the development of modeling and simulation software worldwide, and at least some duplication of effort is apparent. Sharing of models between researchers remains a challenge but is being addressed by the development of several markup languages. Finally, the involvement and interest of industry in use of modeling in biology is significant although, again, not ubiquitous.

REFERENCES

Akutsu, T., S. Miyano and S. Kuhara. 2000. Inferring qualitative relations in genetic networks and metabolic pathways. *Bioinformatics* 16: 727–734.

Albert, R., Y. W. Chiu and H. G. Othmer. 2004. Dynamic receptor team formation can explain the high signal transduction gain in *Escherichia coli.Biophys J* 86: 2650–2659.

Alm, E. and A. P. Arkin. 2003. Biological networks. *Curr Opin Struct Biol* 13: 193–202.

Alon, U., M. G. Surette, N. Barkai and S. Leibler. 1999. Robustness in bacterial chemotaxis. *Nature* 397: 168–171.

Arita, M. 2004. The metabolic world of *Escherichia coli* is not small. *Proc Natl Acad Sci U S A* 101: 1543–1547.

Barabasi, A. L. and E. Bonabeau. 2003. Scale-free networks. *Sci Am* 288: 60–69.

Barabasi, A. L. and Z. N. Oltvai. 2004. Network biology: understanding the cell's functional organization. *Nat Rev Genet* 5: 101–113.

Bhalla, U. S., P. T. Ram and R. Iyengar. 2002. MAP kinase phosphatase as a locus of flexibility in a mitogen-activated protein kinase signaling network. *Science* 297: 1018–1023.

Bolouri, H. and E. H. Davidson. 2003. Transcriptional regulatory cascades in development: initial rates, not steady state, determine network kinetics. *Proc Natl Acad Sci U S A* 100: 9371–9376.

Bray, D. and R. B. Bourret. 1995. Computer analysis of the binding reactions leading to a transmembrane receptor-linked multiprotein complex involved in bacterial chemotaxis. *Mol Biol Cell* 6: 1367–1380.

Carlson, J. M. and J. Doyle. 2002. Complexity and robustness. *Proc Natl Acad Sci U S A* 99 Suppl 1:2538–45.

Carlson, J. M. and J. Doyle. 1999. Highly optimized tolerance: a mechanism for power laws in designed systems. *Phys Rev E. Stat Phys Plasmas. Fluids Relat Interdiscip Topics* 60: 1412–1427.

Chassagnole, C., E. Quentin, D. A. Fell, P. de Atauri, and J. P. Mazat. 2003. Dynamic simulation of pollutant effects on the threonine pathway in *Escherichia coli*. *C R. Biol* 326: 501–508.

Chen, K.C. *et al.* 2004. Integrative analysis of cell cycle control in budding yeast. *Mol Biol Cell* 15: 3841–3862.

Chow, C. C., B. Gutkin, D. Hansel, C. Meunier and J. Dalibard. 2005. Methods and Models in Neurophysics : Proceedings of the Les Houches Summer School 2003 (École D'été de Physique Théoretique, Les Houches//Proceedings). Elsevier Science.

Christopher, R. *et al.* 2004. Data-driven computer simulation of human cancer cell. *Ann NY Acad Sci* 1020:132–53

Crampin, E. J. *et al.* 2004. Computational physiology and the Physiome Project. *Exp Physiol* 89: 1–26.

Cruywagen, G. C., Maini, P. K. & Murray, J. D. 1994. Travelling waves in a tissue interaction model for skin pattern formation. *J Math Biol* 33: 193–210.

Csete, M. E. and Doyle, J. C. 2002. Reverse engineering of biological complexity. *Science* 295: 1664–1669.

Davidson, L. A., M. A. Koehl, R. Keller and G. F. Oster. 1995. How do sea urchins invaginate? Using biomechanics to distinguish between mechanisms of primary invagination. *Development* 121: 2005–2018.

Dayan, P. and L. F. Abbott. 2001. Theoretical Neuroscience: Computational and Mathematical Modeling of Neural Systems. The MIT Press.

Dixit, N. M., J. E. Layden-Almer, T. J. Layden, and A. S. Perelson. 2004. Modeling how ribavirin improves interferon response rates in hepatitis C virus infection. *Nature* 432: 922–924.

Doi, A., S. Fujita, H. Matsuno, M. Nagasaki and S. Miyano. 2004. Constructing biological pathway models with hybrid functional Petri nets. *In Silico Biol* 4: 271–291.

Ebenhoh, O., T. Handorf and R. Heinrich. 2004. Structural analysis of expanding metabolic networks. *Genome Inform Ser Workshop Genome Inform* 15: 35–45.

Erban, R. and H. G. Othmer. 2004. From individual to collective behavior in bacterial chemotaxis. *SIAM J Appl Math* 65(2): 361–391.

Garny, A., P. Kohl, P. J. Hunter, M. R. Boyett and D. Noble. 2003. One-dimensional rabbit sinoatrial node models: benefits and limitations. *J Cardiovasc Electrophysiol* 14: S121-S132.

Gilchrist, M. A., D. Coombs, and A. S. Perelson. Optimizing within-host viral fitness: infected cell lifespan and virion production rate. *J Theor Biol* 229: 281–288.

Goldbeter, A. 2002. Computational approaches to cellular rhythms. *Nature* 420: 238–245.

Goldbeter, A. *et al.* 2001. From simple to complex oscillatory behavior in metabolic and genetic control networks. *Chaos* 11: 247–260.

Hodgkin, A. L. and A. F. Huxley. 1952. A quantitative description of membrane current and its application to conduction and excitation in nerve. *J Physiol* 117: 500–544.

Hucka, M. *et al.* 2003. The systems biology markup language (SBML): a medium for representation and exchange of biochemical network models. *Bioinformatics* 19: 524–531.

Igoshin, O. A., R. Welch, D. Kaiser and G. Oster. 2004. Waves and aggregation patterns in myxobacteria. *Proc Natl Acad Sci U S A* 101: 4256–4261.

Jeong, H., B. Tombor, R. Albert, Z. N. Oltvai and A. L. Barabasi. 2000. The large-scale organization of metabolic networks. *Nature* 407: 651–654.

Kaazempur-Mofrad, M. R. *et al.* 2004. Characterization of the atherosclerotic carotid bifurcation using MRI, finite element modeling, and histology. *Ann Biomed Eng* 32: 932–946.

Kitano, H. 2004. Biological robustness. *Nat Rev Genet* 5: 826–837.

Kitano, H. *et al.* 2004. Metabolic syndrome and robustness tradeoffs. *Diabetes* 53: Suppl 3:S6-S15.

Koch, C. 2004. Biophysics of Computation: Information Processing In Single Neurons. Oxford University Press.

Koch, C. and I. Segev. 1998. Methods in Neuronal Modeling: From Ions to Networks. The MIT Press.

Kremling, A. *et al.* 2004. A benchmark for methods in reverse engineering and model discrimination: problem formulation and solutions. *Genome Res* 14: 1773–1785.

Itzkovitz, S. and U. Alon. 2005. Subgraphs and network motifs in geometric networks. *Phys Rev E Stat Nonlin Soft Matter Phys* 71: 026117.

Lee, E., A. Salic, R. Kruger, R. Heinrich and M. W. Kirschner. 2003. The roles of APC and Axin derived from experimental and theoretical analysis of the Wnt pathway. *PLoS Biol* 1: E10.

Leloup, J. C. & A. Goldbeter. 2003. Toward a detailed computational model for the mammalian circadian clock. *Proc Natl Acad Sci U S A* 100: 7051–7056.

Lipkow, K., S. S. Andrews and D. Bray. 2005. Simulated diffusion of phosphorylated CheY through the cytoplasm of *Escherichia coli*. *J Bacteriol* 187: 45–53.

Luo, C. H. and Y. Rudy. 1994. A dynamic model of the cardiac ventricular action potential. I. Simulations of ionic currents and concentration changes. *Circ Res* 74: 1071–1096.

Luo, C. H. and Y. Rudy. 1994. A dynamic model of the cardiac ventricular action potential. II. Afterdepolarizations, triggered activity, and potentiation. *Circ Res* 74: 1097–1113.

Marino, S. and D. E. Kirschner. 2004. The human immune response to Mycobacterium tuberculosis in lung and lymph node. *J Theor Biol* 227: 463–486.

Markhasin, V. S. *et al.* 2003. Mechano-electric interactions in heterogeneous myocardium: development of fundamental experimental and theoretical models. *Prog Biophys Mol Biol* 82: 207–220.

Matsuoka, S., N. Sarai, H. Jo and A. Noma. 2004. Simulation of ATP metabolism in cardiac excitation-contraction coupling. *Prog Biophys Mol Biol* 85: 279–299.

McCulloch, A. D., P. J. Hunter, and B. H. Smaill. 1992. Mechanical effects of coronary perfusion in the passive canine left ventricle. *Am J Physiol* 262: H523-H530.

McGee, P. 2005. Modeling Success with *In Silico* Tools. *Drug Discovery and Development* 8(4): 24–28.

Morohashi, M. *et al.* 2002. Robustness as a measure of plausibility in models of biochemical networks. *J Theor Biol* 216: 19–30.

Noble, D. 2002. Modeling the heart: insights, failures and progress. *Bioessays* 24: 1155–1163.

Novak, B. and J. J. Tyson. 2003. Modeling the controls of the eukaryotic cell cycle. *Biochem Soc Trans* 31: 1526–1529.

Park, C. S., I. C. Schneider and J. M. Haugh. 2003. Kinetic analysis of platelet-derived growth factor receptor/phosphoinositide 3-kinase/Akt signaling in fibroblasts. *J Biol Chem* 278: 37064–37072.

Patnaik, R. and J. C. Liao. 1994. Engineering of *Escherichia coli* central metabolism for aromatic metabolite production with near theoretical yield. *Appl Environ Microbiol* 60: 3903–3908.

Patnaik, R. and R. G. L. J. C. Spitzer. 1995. Pathway Engineering for Production of Aromatics in *Escherichia coli*: Confirmation of Stoichiometric Analysis by Independent Modulation of AroG, TktA, and Pps activities. *Biotech Bioeng* 46: 361–370.

Peirce, S. M., E. J. Van Gieson and T. C. Skalak. 2004. Multicellular simulation predicts microvascular patterning and *in silico* tissue assembly. *FASEB J* 18: 731–733.
Poolman, M. G., H. E. Assmus and D. A. Fell. 2004. Applications of metabolic modeling to plant metabolism. *J Exp Bot* 55: 1177–1186.
Pribyl, M., C. B. Muratov and S. Y. Shvartsman. 2003. Discrete models of autocrine cell communication in epithelial layers. *Biophys J* 84, 3624–3635.
Pribyl, M., C. B. Muratov and S. Y. Shvartsman. 2003. Transitions in the model of epithelial patterning. *Dev Dyn* 226: 155–159.
Ramanujan, S., G. C. Koenig, T. P. Padera, B. R. Stoll, and R. K. Jain. 2000. Local imbalance of proangiogenic and antiangiogenic factors: a potential mechanism of focal necrosis and dormancy in tumors. *Cancer Res* 60: 1442–1448.
Ramsey, S., D. Orrell, and H. Bolouri. 2005. Dizzy: stochastic simulation of large-scale genetic regulatory networks. *J Bioinform Comput Biol* 3: 415–436.
Sarkar, C. A. *et al.* 2002. Rational cytokine design for increased lifetime and enhanced potency using pH-activated "histidine switching." *Nat Biotechnol* 20: 908–913.
Sarkar, C. A. and D. A. Lauffenburger. 2003. Cell-level pharmacokinetic model of granulocyte colony-stimulating factor: implications for ligand lifetime and potency *in vivo*. *Mol Pharmacol* 63: 147–158.
Savoie, C. J. *et al.* 2003. Use of gene networks from full genome microarray libraries to identify functionally relevant drug-affected genes and gene regulation cascades. *DNA Res* 10: 19–25.
Schafer, J. R., D. A. Fell, D. Rothman and R. G. Shulman. 2004. Protein phosphorylation can regulate metabolite concentrations rather than control flux: the example of glycogen synthase. *Proc Natl Acad Sci U S A* 101: 1485–1490.
Schmid, J. W., K. Mauch, M. Reuss, E. D. Gilles, and A. Kremling. 2004. Metabolic design based on a coupled gene expression-metabolic network model of tryptophan production in *Escherichia coli*. *Metab Eng* 6: 364–377.
Schoeberl, B., U. B. Nielsen, D. A. Lauffenburger, and P. K. Sorger. 2003. Network topology and distinct protein expression levels: enough to predict signal transduction *in silico*? Proceedings of the International Congress of Systems Biology, 64–65.
Segal, E. *et al.* 2003. Module networks: identifying regulatory modules and their condition-specific regulators from gene expression data. *Nat Genet.* 34: 166–176.
Shimizu, T. S., S. V. Aksenov, and D. Bray 2003. A spatially extended stochastic model of the bacterial chemotaxis signalling pathway. *J Mol Biol* 329: 291–309.
Stelling, J. and E. D. Gilles. 2004. Mathematical modeling of complex regulatory networks. *IEEE Trans Nanobioscience* 3: 172–179.
Stelling, J., E. D. Gilles, and F. J. Doyle III. 2004. Robustness properties of circadian clock architectures. *Proc Natl Acad Sci U S A* 101: 13210–13215.
Stelling, J., S. Klamt, K. Bettenbrock, S. Schuster, and E. D. Gilles. 2002. Metabolic network structure determines key aspects of functionality and regulation. *Nature* 420: 190–193.
Stoll, B. R., C. Migliorini, A. Kadambi, L. L. Munn, and R. K. Jain. 2003. A mathematical model of the contribution of endothelial progenitor cells to angiogenesis in tumors: implications for antiangiogenic therapy. *Blood* 102: 2555–2561.
Sveiczer, A., J. J. Tyson, and B. Novak. 2004. Modeling the fission yeast cell cycle. *Brief Funct Genomic Proteomic* 2: 298–307.
Swameye, I., T. G. Muller, J. Timmer, O. Sandra, and U. Klingmüller. 2003. Identification of nucleocytoplasmic cycling as a remote sensor in cellular signaling by databased modeling. *Proc Natl Acad Sci U S A* 100: 1028–1033.
ten Tusscher, K. H., D. Noble, P. J. Noble, and A. V. Panfilov. 2004. A model for human ventricular tissue. *Am J Physiol Heart Circ Physiol* 286: H1573-H1589.

Thomas, S., P. J. Mooney, M. M. Burrell, and D. A. Fell. 1997. Metabolic control analysis of glycolysis in tuber tissue of potato (*Solanum tuberosum*): explanation for the low control coefficient of phosphofructokinase over respiratory flux. *Biochem J* 322: 119–127.

Tranquillo, R. T. and J. D. Murray. 1993. Mechanistic model of wound contraction. *J Surg Res* 55: 233–247.

Trimmer, J., C. McKenna, B. Sudbeck, and R. Ho. 2005. Use of Systems Biology in Clinical Development: Design and Prediction of a Type 2 Diabetes Clinical Trial. PAREXEL Pharmaceutical R&D Sourcebook 2004/2005, 131–132.

von Dassow, G., E. Meir, E. M. Munro and G. M. Odell. 2000. The segment polarity network is a robust developmental module. *Nature* 406: 188–192.

von Dassow, G. and G. M. Odell. 2002. Design and constraints of the Drosophila segment polarity module: robust spatial patterning emerges from intertwined cell state switches. *J Exp Zool* 294: 179–215.

Winslow, R.L. *et al.* 2000. Electrophysiological modeling of cardiac ventricular function: from cell to organ. *Annu Rev Biomed Eng* 2:119–55.

Woolf, P.J. and J. J. Linderman. 2003. Untangling ligand induced activation and desensitization of G-protein-coupled receptors. *Biophys J* 84: 3–13.

Wuchty, S., Z. N. Oltvai and A. L. Barabasi. 2003. Evolutionary conservation of motif constituents in the yeast protein interaction network. *Nat Genet* 35: 176–179.

Yi, T. M., H. Kitano and M. I. Simon. 2003. A quantitative characterization of the yeast heterotrimeric G protein cycle. *Proc Natl Acad Sci U S A* 100: 10764–10769.

Yook, S. H., H. Jeong and A. L. Barabasi. 2002. Modeling the Internet's large-scale topology. *Proc Natl Acad Sci U S A* 99: 13382–13386.

Zwolak, J. W., J. J. Tyson, and L. T. Watson. 2005. Parameter estimation for a mathematical model of the cell cycle in frog eggs. *J Comput Biol* 12: 48–63.

CHAPTER 5

Systems Biology in Plant Research

Fumiaki Katagiri

INTRODUCTION

Generally speaking, systems biology research in plants has not reached an advanced stage, a situation largely due to the low level of funding for basic plant research. There are two major reasons for this. First, crop species to which plant research outcomes could be applied are numerous and diverse, and many problems that need to be solved are species-specific. Therefore, resources are spread thinly among many different plant systems. There is also strong political pressure to shift plant research funding toward crop species at the expense of model systems, even though the crop systems are generally less tractable. Since quality, quantity, and correlativity of experimental data and a repeating cycle of experimental and theoretical work are critical for success in systems biology, it would be best to focus on model plant species at this stage. Second, the agribusiness industry is showing declining interest in applications of biotechnology. Due to the anti-GMO (genetically modified organism) movement in developed countries, the industry expects difficulty in marketing biotech-based high-value products, such as functional foods, which could involve complex metabolic engineering. The declining research activities in the agribusiness industry also reduce the employment prospects for scientists trained in plant biology. Low funding levels are directly related to slower progress in new basic research areas, such as systems biology, especially when the research requires relatively large early investments and long periods of time for outcomes to be apparent. Furthermore, interdisciplinary research areas such as systems biology need to attract researchers from different disciplines to particular biological systems, which is difficult when funding levels are low and future job prospects are poor. In this context, it was

M. Cassman et al. (eds.), Systems Biology, 83–90.

disappointing to learn that a major plant systems biology collaborative program in Japan, the rice genome simulator project, was abruptly cancelled last year without clear explanation.

Another challenge, which is not specific to the plant research field, is that the number of senior principal investigators (PIs) who can properly evaluate systems biology projects is very small. Senior PIs are the ones who would organize large projects, make major changes in curricula, review manuscripts, evaluate grant proposals, and promote young researchers. Due to the enormous success of molecular genetics for many years, a large percentage of senior PIs cannot appreciate projects that are not reductionism-oriented and do not yield clear yes/no answers based on highly simplified hypotheses. The effect of this on systems biology funding is severe. This bias constitutes a drag on the adoption of systems biology by the research community. Training the next generation of researchers in systems biology is another important issue, and it is discussed in detail in Chapter 6. However, this senior PI issue is, in a sense, more important than training young researchers. Young researchers would easily be discouraged if they were not properly evaluated and encouraged by people with the power to influence their careers.

To accurately model a biological network, researchers need to know the identity and function of a sufficient number of molecular components that correspond to nodes in the network, although what fraction is sufficient is frequently debated. Molecular networks in many plants are generally less well-studied than in animals and microbes, and it is believed by many that not enough is yet known to move to modeling. Therefore, in many networks, identification of components is a major task at present. Functional genomics, i.e. discovery of gene function on a large scale, is a popular approach for this purpose. Public funding, such as *Arabidopsis* 2010 by the National Science Foundation (NSF) and programs funded by the EU, is supporting many ongoing functional genomics efforts in the U.S., Europe, and Japan. The panel visited some of the major sites for these efforts, such as Dr. Beynon's group for the Complete *Arabidopsis* Transcriptome MicroArray (CATMA) program (http://www.catma.org) in Europe (Hilson et al., 2004) and Dr. Shinozaki's group at the RIKEN Genome Sciences Center in Japan (Sakurai et al., 2005). However, plant research programs that have advanced beyond the stage of component identification do exist. In the following sections, some examples of advanced plant systems biology research are reviewed. The sections are divided according to the areas of research rather than by geographic areas. This is because clear regional differences were not apparent and because it is crucial for the plant research community to coordinate and cooperate at the global level since the total funds available for research in these areas are limited.

METABOLIC NETWORKS

The structures of major metabolic networks are well established. Some groups are already studying dynamics of the networks. While metabolic networks can be illustrated as conventional metabolic maps, it is not clear which routes are really important to explain the fluxes in these networks. Dr. Fell (Oxford Brookes University) applied the elementary mode analysis (Schuster et al., 2002) to phase 3 (daylight metabolism of stored malate with no net CO_2 uptake) of the Crassulacean acid metabolism (CAM). CAM plants are typically plants of arid climates. They open stomata during the night to assimilate CO_2 into malate, and during the day they use the carbon stored in malate for Rubisco (ribulose 1,5-bisphosphate carboxylase/oxygenase) -driven carbon fixation without opening stomata. In this way they minimize loss of water. The analysis predicted six distinct pathways that could be used to accomplish CAM whereas only five pathways were known. Later, *Mesembryanthemum crystallinum* (common ice plant) was found to use the sixth pathway for CAM, validating the prediction.

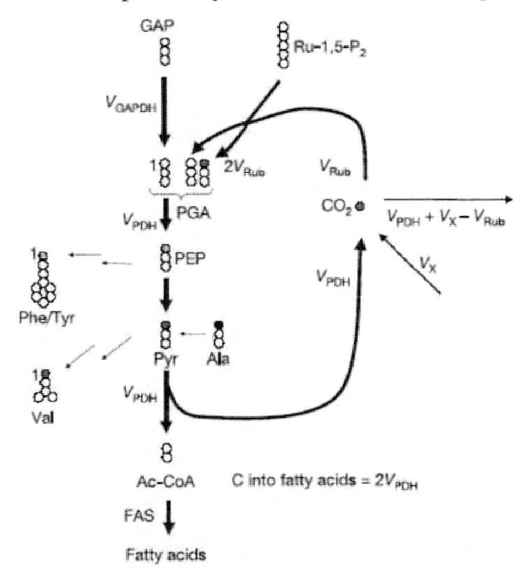

Figure 5.1. A newly discovered pathway for refixation of CO_2 released during conversion of carbohydrate to storage fat (Schwender et al., 2004).

Dr. Shachar-Hill's group (Michigan State University) combined the Elementary Mode Analysis with experimental measurements of mass balance, enzyme activity and stable isotope labeling in a study of carbohydrate conversion to oil through glycolysis in immature green seeds of *Brassica napus* (oilseed rape) (Schwender et al., 2004). The efficiency of conversion was higher than expected. They discovered that refixation by Rubisco of released CO_2 explains the high efficiency. This was the first description of the role of Rubisco in this context.

Major efforts to collect correlated messenger ribonucleic acid (mRNA) profiles, metabolite profiles, and other phenotypes from many genetically perturbed *Arabidopsis* plants are under way at the Max Planck Institute for Molecular Plant Physiology (http://www.mpimp-golm.mpg.de/) in Golm, Germany and by members of a collaboration led by the Kazusa DNA Research Institute (KDRI) (http://www.kazusa.or.jp/eng/index.html) in Chiba, Japan. The research in Germany is led by Drs. Willmitzer and Stitt and focuses on primary metabolism. In addition to use of the Affymetrix GeneChip array, they collect mRNA profiles of most transcription factor genes using a high-throughput real-time reverse transcription-polymerase chain reaction (RT-PCR) (Czechowski et al., 2004). They also measure protein levels of major enzymes and use other specialized profiling methods. The research in Japan is led by Drs. Shibata (KDRI) and Saito (Chiba University) and focuses on secondary metabolism. They use sophisticated equipment (gas chromatography time-of-flight mass spectrometry—GC-TOF-MS, Liquid chromatography photo-diode-array-detection mass spectrometry—LC-PDA-MS, Capillary electrophoresis mass spectrometry—CE-MS, liquid chromatography fourier transfor-mation mass spectrometry—LC-FT-MS, and liquid chromatography time-of-flight mass spectrometry—LC-TOF-MS) for metabolomic meas-urements in addition to mRNA profiling with the Agilent microarray (Hirai et al., 2004). Accurate mass information obtained by Fourier transformation mass spectrometry (FT-MS) is a great help in identification of metabolites. The Japanese group uses a well-established suspension culture cell line to ease the issue of establishing consistent growth conditions. Both German and Japanese groups have invested in bioinformatic tools, including viewers that integrate expression and metabolite information along metabolic maps (Thimm et al., 2004).

To understand the dynamics of metabolic networks, metabolic flux measurements are important. Dr. Shanks (Iowa State University) has developed a computer-assisted method to estimate metabolic fluxes of several pathways using biosynthetically directed fractional ^{13}C-labeling and two-dimensional [^{13}C, ^{1}H] nuclear magnetic resonance (NMR) (Sriram et al., 2004). This method could be applicable for many known metabolic networks.

REGULATORY NETWORKS IN DEVELOPMENTAL PROCESSES

Thanks to extensive genetic analysis, many important components of regulatory networks that control several developmental processes are known. If key information can be collected about the important components, sufficient information may be obtained to model such regulatory networks at a practical level. Some developmental processes are explained by transcription-regulatory cascades. In such cases, the activity of each gene is well correlated with the mRNA level of the gene. For plants to be successful in evolutionary terms, it is crucial to have flowers open and seeds set at the right time in the growing season. Many factors, such as photoperiod (day length), the plant hormones called gibberellins, and experience of cold weather, can affect flowering time. Dr. Welch (Kansas State University) modeled the *Arabidopsis* flowering time control system using mRNA levels of the important molecular components of the process (Welch et al., 2003). Circadian rhythms are important in developmental controls as well as in physiological controls. For example, plants measure photoperiod by comparing it with their own circadian rhythms. Dr. Millar and his collaborators (University of Warwick, U.K.) modeled the *Arabidopsis* circadian clock (Locke et al., 2005). Although circadian clocks exist in diverse organisms, such as cyanobacteria, fungi, animals, and plants, the molecular machinery in different organisms seems to be quite diverse. In the case of cyanobacteria, an *in vitro* system composed of three proteins that can generate a 24-hour period has recently been reconstituted (Nakajima et al., 2005). It is of interest to see whether the network structures and the system control are also diverse, even though the resulting clocks are all robust in the maintaining period.

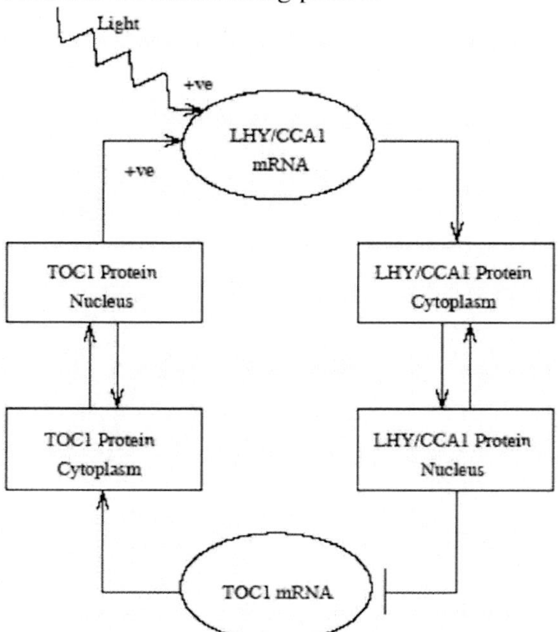

Figure 5.2. Model for the central feedback in the *Arabidopsis* clock (Locke et al., 2005).

An ambitious project called The Computable Plant has been initiated by a consortium led by Dr. Meyerowitz (Cal Tech) (http://www. computableplant.org/). This project aims to model development of the shoot apical meristem (SAM) in *Arabidopsis*. Meristems are the inner plant tissues, where regulated cell division, pattern formation, and differentiation give rise to plant parts like leaves and flowers. The project includes modeling efforts as well as experimental efforts to monitor the cell lineage of specific cell types in real time (Reddy et al., 2004). This process will ultimately be automated.

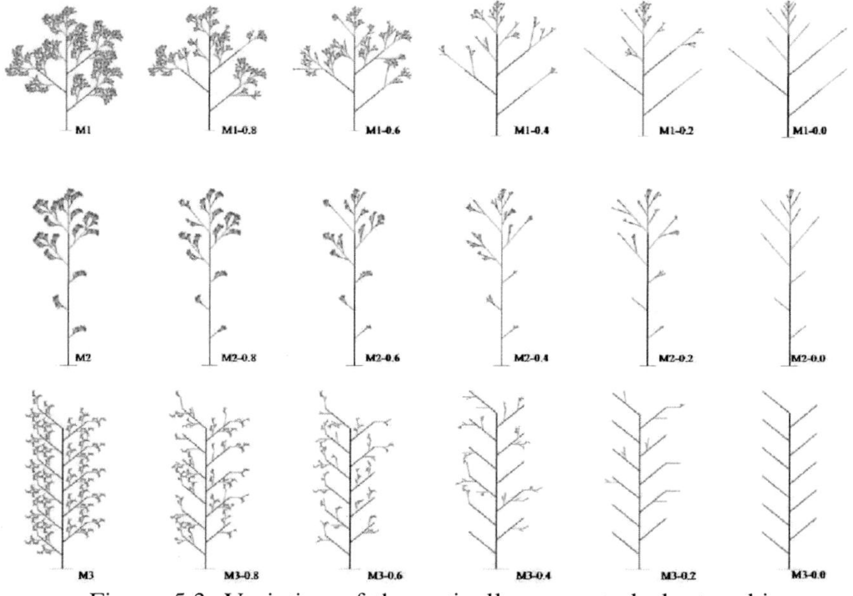

Figure 5.3. Variation of theoretically generated plant architectures (Ferraro et al., 2005).

Morphogenesis has been a central interest in developmental biology. Some modeling work has been done to explain plant morphogenesis at various scales. Dr. Coen's group (John Innes Center, Norwich, U.K.) measured cell growth in the snapdragon petal using clonal analysis (Rolland-Lagan et al., 2003). Then they modeled the complex shape of the petal based on the cell growth pattern. Simulation of the model found that the direction of growth that was maintained parallel to the proximodistal axis of the flower was crucial for formation of the asymmetric petal shape, but other factors, such as changes in cell shape, were not. Dr. Prusinkiewicz's work (University. of Calgary, Canada) involves mathematical modeling of the plant architecture using L-systems with a small number of parameters (Prusinkiewicz, 2004). For example, diverse inflorescence shape patterns seen among various dicot plant species could be explained by two parameters that represent characteristics of a putative factor. This approach can be considered as reverse engineering of the plant architecture. It will be interesting to see if an actual molecular counterpart of the putative factor exists.

CONCLUDING REMARKS

Progress of systems biology research in the plant field has been slow. However, several advanced studies shed light on unique aspects of plants. There are no clear national differences in plant research in systems biology. Several actions are needed to promote systems biology of plants.

To make the most out of limited funding:

- Focus on model plant species. It is clear that the majority of advanced studies have been performed with model plant species, such as *Arabidopsis*.
- Cooperate rather than compete at the global level.

To compensate for the PI population bias against promotion of systems biology in the research community:

- Implement a sustaining, targeted funding program in plant systems biology.

To raise the next generation of researchers:

- Train biology-major students in quantitative science.
- Recruit students oriented to mathematics, engineering, physics, and chemistry into plant biology.

REFERENCES

Czechowski, T., R. P. Bari, M. Stitt, W. R. Scheible, and M. K. Udvardi. 2004. Real-time RT-PCR profiling of over 1400 *Arabidopsis* transcription factors: unprecedented sensitivity reveals novel root- and shoot-specific genes. *Plant J* 38: 366–379.

Ferraro, P. C. Godin, and P. Prusinkiewicz. 2005. Toward a quantification of self-similarity in plants. *Fractals* 13(2): 91–109.

Hilson, P., J. Allemeersch, T. Altmann, S. Aubourg, A. Avon, J. Beynon, R. P. Bhalerao, F. Bitton, M. Caboche, B. Cannoot, *et al.* 2004. Versatile gene-specific sequence tags for *Arabidopsis* functional genomics: transcript profiling and reverse genetics applications. *Genome Res* 14: 2176–2189.

Hirai, M. Y., M. Yano, D. B. Goodenowe, S. Kanaya, T. Kimura, M. Awazuhara, M. Arita, T. Fujiwara, and K. Saito. 2004. Integration of transcriptomics and metabolomics for understanding of global responses to nutritional stresses in *Arabidopsis* thaliana. *Proc Natl Acad Sci U S A* 101: 10205–10210.

Locke, J. C., A. J. Millar, and M. S. Turner. 2005. Modeling genetic networks with noisy and varied experimental data: the circadian clock in *Arabidopsis* thaliana. *J Theor Biol* 234: 383–393.

Nakajima, M., K. Imai, H. Ito, T. Nishiwaki, Y. Murayama, H. Iwasaki, T. Oyama, and T. Kondo. 2005. Reconstitution of circadian oscillation of cyanobacterial KaiC phosphorylation *in vitro*. *Science* 308: 414–415.

Prusinkiewicz, P. 2004. Modeling plant growth and development. *Curr Opin Plant Biol* 7: 79–83.

Reddy, G. V., M. G. Heisler, D. W. Ehrhardt and E. M. Meyerowitz. 2004. Real-time lineage analysis reveals oriented cell divisions associated with morphogenesis at the shoot apex of *Arabidopsis thaliana*. *Development* 131: 4225–4237.

Rolland-Lagan, A. G., J. A. Bangham, and E. Coen. 2003. Growth dynamics underlying petal shape and asymmetry. *Nature* 422: 161–163.

Sakurai, T., Satou, M., Akiyama, K., Iida, K., Seki, M., Kuromori, T., Ito, T., Konagaya, A., Toyoda, T., and Shinozaki, K. 2005. RARGE: a large-scale database of RIKEN *Arabidopsis* resources ranging from transcriptome to phenome. *Nucleic Acids Res* 33: D647–650.

Schuster, S., C. Hilgetag, J. H. Woods, and D. A. Fell. 2002. Reaction routes in biochemical reaction systems: algebraic properties, validated calculation procedure and example from nucleotide metabolism. *J Math Biol* 45: 153–181.

Schwender, J., F. Goffman, J. B. Ohlrogge, and Y. Shachar-Hill. 2004. Rubisco without the Calvin cycle improves the carbon efficiency of developing green seeds. *Nature* 432: 779–782.

Sriram, G., D. B. Fulton, V. V. Iyer, J. M. Peterson, R. Zhou, M. E. Westgate, M. H. Spalding, J. V. and Shanks. 2004. Quantification of compartmented metabolic fluxes in developing soybean embryos by employing biosynthetically directed fractional (13)C labeling, two-dimensional [(13)C, (1)H] nuclear magnetic resonance, and comprehensive isotopomer balancing. *Plant Physiol* 136: 3043–3057.

Thimm, O., O. Blasing, Y. Gibon, A. Nagel, S. Meyer, P. Kruger, J. Selbig, L. A. Muller, S. Y. Rhee, and M. Stitt. 2004. MAPMAN: a user-driven tool to display genomics data sets onto diagrams of metabolic pathways and other biological processes. *Plant J* 37: 914–939.

Welch, S. M., J. L. Roe, and Z. Dong. 2003. A genetic neural network model of flowering time control in *Arabidopsis thaliana*. *Agron J* 95: 71–81.

CHAPTER 6

Education, National Programs, and Infrastructure in Systems Biology

Marvin Cassman, Douglas Lauffenburger, and Frank Doyle

INTRODUCTION

The future of systems biology will depend on three critical elements: education of a new generation of scientists who have both biological and mathematical training; the availability of funding that operates outside of disciplinary boundaries; and the availability of a supportive infrastructure that can accommodate the needs of an intrinsically interdisciplinary research area. This chapter will consider the status of these issues both in the U.S. and elsewhere.

EDUCATION

Not surprisingly for such a hot field, systems biology has spurred interest from thousands of researchers, some just starting their careers, others well established but looking for an opportunity to become involved. Because of the need to couple computational analysis techniques with systematic biological experimentation, more and more universities are offering PhD programs that integrate both computational and biological subject matter.

Page 62 lists a set of current educational programs in this field. Several of the early endeavors in the U.S. are closely associated with bioengineering. The Computational & Systems Biology graduate program was established at the Massachusetts Institute of Technology (MIT) in 2004 as a three-way partnership among the biological engineering, biology, and electrical engineering and computer science departments. The University of California, San Diego (UCSD) offers a systems biology track within its

M. Cassman et al. (eds.), Systems Biology, 91–111.

Bioengineering graduate program. Other nascent programs are not closely associated with formal engineering programs but instead arise out of life sciences or medical sciences fields. Examples of these include the new programs at Harvard Medical School, the Institute for Systems Biology, Oxford University, and Biocentrum Amsterdam.

Given the pace of the field, it is certainly too early to endorse a particular syllabus as the correct or best option. However, the study of systems biology must lead to a rigorous understanding of both experimental biology and quantitative modeling. Programs might require that all students, regardless of background, perform hands-on research in both computer programming and in the wet laboratory. Required coursework in biology typically includes genetics, biochemistry, molecular and cell biology, with lab work associated with each of these. Coursework in quantitative modeling might include probability, statistics, information theory, numerical optimization, artificial intelligence and machine learning, graph and network theory, and nonlinear dynamics. Of the biological coursework, genetics is particularly important, because the logic of genetics is, to a large degree, the logic of systems biology. Of the coursework in quantitative modeling, graph theory and machine-learning techniques are of particular interest, because systems approaches often reduce cellular function to a search on a network of biological components and interactions. A course of study integrating life and quantitative sciences helps students to appreciate the practical constraints imposed by experimental biology and to effectively tailor research to the needs of the laboratory biologist. At the same time, knowledge of the major algorithmic techniques for analysis of biological systems will be crucial for making sense of the data.

An alternative to pursuing a cross-disciplinary program is to tackle one field initially and then learn another in graduate school. Examples would include choosing an undergraduate major in bioengineering and then obtaining a PhD in molecular biology, or starting within biochemistry then pursuing graduate coursework in bioengineering and systems biology. This leads to a common question: when contemplating a transition, is it better to switch from quantitative sciences to biology or vice versa? Although some believe that it is easier to move from engineering into biology, the honest answer is that either trajectory can work. Some practical advice is that if coming from biology, it is best to start by becoming familiar with Unix, Perl, and Java before diving into more complex computational methodologies. If coming from the quantitative sciences, an effective strategy is to jump into a wet laboratory as soon as possible.

With the formation of myriad new academic departments and centers, the academic job market is booming. On the other hand, biotechnology firms and 'big pharma' have been more cautious about getting involved. However, most agree that in the long term systems approaches promise to influence drug development in several areas: (a) target identification, in which drugs are developed to target a specific molecule or molecular interaction within a pathway; (b) prediction of drug mechanism of action (MOA), in which a compound has known therapeutic effects but the molecular mechanisms by which it achieves these effects are unclear; and (c) prediction of drug toxicity and properties related to absorption, distribution, metabolism, and excretion (ADME/Tox). In all of these cases, the key contribution of systems biology would be a comprehensive blueprint of cellular pathways used for identifying proteins at key pathway control points, or proteins for which the predicted perturbation phenotypes most closely resemble those observed experimentally with a pharmacologic or toxic agent.

Looking toward higher levels of living systems behavioral hierarchy, students preparing for research careers in integrative systems physiology should build a strong foundation in core life sciences, mathematics and engineering. It is particularly useful to be immersed in life sciences courses which present biological principles in the context of mathematical models and engineering methodologies. An example of such a course is the year-long course entitled Physiological Foundations of Biomedical Engineering offered in the Biomedical Engineering department at the Johns Hopkins University. Foundation courses in mathematics could include ordinary and partial differential equation theory as well as probability theory and stochastic processes. While not commonly available, introductory course work in nonlinear dynamical systems theory would be valuable. Students may also opt to build a strong foundation in a core engineering discipline such as mechanical, chemical or electrical engineering.

Students pursuing any aspect of computational or systems biology at the graduate level face the hard fact that they *must* be as deeply educated in relevant areas of the life sciences as their biological colleagues, and they *must* be as strong in appropriate areas of engineering and mathematics as their colleagues in traditional areas of engineering and mathematics. Students will only be successful in this endeavor if they have a true love for both their chosen areas of biology and math/engineering. The broad discipline of quantitative modeling of biological systems is one that is developing rapidly and is seeing increased representation in bio- and biomedical engineering, life sciences and traditional engineering departments. Students may therefore undertake combined experimental and modeling research or modeling research conducted in collaboration with experimental investigators with reasonable confidence that they will be able to find an academic department which appreciates and supports the particular balance they have chosen between modeling and experimentation.

Comparative Programs

A comparison between U.S. and other countries in education is difficult, largely because the programs that do exist are of such recent origin. One striking difference is the major role that engineering has played in the development of systems biology and systems biology training in the U.S. and its relative absence in the EU and the U.K. (This was not as true in Japan and clear exceptions can be found in Europe, such as the Max Planck Institute for Complex Technical Systems in Magdeburg.) In general, the backgrounds of the major practitioners outside the U.S. show a heavier representation in physics and mathematics (the latter particularly in the U.K.) and very little in engineering. This is stated neither as a positive or negative, but it does reflect organizational and cultural differences that will be inevitably be reflected in the approaches to training.

However, at this point in time the most identifiable characteristic of training in systems biology is its absence. The most common response to queries about educational programs was that they didn't exist. Since the research efforts are themselves of recent origin perhaps this isn't surprising. However, it is a significant barrier to the development of the field. It is therefore worth examining some of the programs that the WTEC panel was able to see.

Japan

Department of Computational Biology, Graduate School of Frontier Sciences, University of Tokyo

This is the first official department of computational biology in Japan, although others are emerging. It began in 2003 and has six faculty and about 30 graduate students. Although its focus is on bioinformatics and "omics" rather than systems biology, it is clearly moving in the direction of modeling and analysis of dynamics in biological systems.

Bioinformatics Center, Institute for Chemical Research, Kyoto University

Human Genome Center, Institute of Medical Science, University of Tokyo

This is a joint effort that has both teaching faculties and has generated a curriculum in bioinformatics that is widely used. Although most of the courses are focused on genomics, there are also courses on network analysis and pathway reconstruction, and modeling and simulation.

Institute for Advanced Biosciences, Keio University

This is both a graduate and undergraduate program with a wide array of courses from genomics to genetic networks to software engineering. A striking innovation is a student laboratory where bioinformatics students (presumably those without much of a background in biology) are trained in experimental techniques for up to a year.

Europe

Humboldt University, Berlin

Humboldt University has an Institute for Theoretical Biology, a Department of Theoretical Biophysics, and a graduate school in the Dynamics of Cellular processes. Courses in modeling began as long ago as 1993, making it an early adopter in areas related to systems biology. Students come from all of these programs to take a well-developed curriculum in bioinformatics, theoretical biology, and biophysics including formal courses on systems biology and mathematical modeling. They also run two-week workshops in the area. International graduate collaboration programs have been established with Bioinformatics at Boston University, the Kyoto Genomics and Bioinformatics Center, and with the BioCentrum in Amsterdam. These programs include joint workshops, PhD student exchanges and post-doctoral fellowships.

Free University of Brussels

The Free University is initiating a MS degree in Bioinformatics and Modeling. This will have three orientations, one of which would be chosen by the students during the second year of the MS: "classical" bioinformatics; computational structural biology; and modeling of dynamic biological phenomena. This will not begin until 2007.

Free University of Amsterdam

The Free University has a new MS program in Biomolecular Integration/Systems Biology. The aim is to provide both expertise in advanced conceptual and modeling methodologies as well as insight into important biological/biomedical issues. It is a two-year program, and involves a detailed research project where the student spends half the time in Amsterdam and half the time with a partner group in a different country, both locations being involved in the advising. Teaching efforts at the PhD level were still evolving at the time of our visit, and involved a joint graduate school with Humboldt University.

Centre for Mathematics and Physics in the Life Sciences, University College, London

This provides a PhD program that trains students who arrive with a background in biology in mathematics, while students with a background in physics and math take courses in biology. All take a course in modeling and bioinformatics and one in physical techniques in the life sciences. This is followed by a set of case studies in interdisciplinary research plus seminars and special courses. The didactic part of the program is one year, followed by three years of research.

University of Warwick, Mathematics Institute

Interdisciplinary Program in Cellular Regulation
Molecular Organization and Assembly in Cells

This is a four-year program which currently aims to train eight postdocs with backgrounds in mathematics so that they are equipped to study biological problems. The program is focused on theoretical analysis, and the funding does not support experimental data generation. Postdocs are trained in biology through seminars, journal clubs, and attendance in group meetings of biology labs and single-afternoon symposia designed for the program. They take courses for MOAC (Molecular Organization and Assembly in Cells) students (see below). They are also paired with biologists for particular projects. The PhD program in MOAC integrates areas of mathematics, biology, and chemistry. This program is funded by a seven-year Doctoral Training Center grant which started in 2003. The program starts with a six-month course for the three areas, followed by a lab rotation in each of the three areas before students choose labs for their PhD projects. Most students entering the program usually have backgrounds in physical chemistry.

Conclusions

The general impression is that most of the formal teaching programs, in the U.S. and abroad, are in bioinformatics rather than systems biology. Relatively few examples of training in modeling are focused on biological systems, and where they do exist they tend to be isolated courses rather than fully integrated programs in systems biology. Most of the programs offer somewhat *ad hoc* "menu selection" curricula. The difficulty of training quantitative students in biology and *vice versa* is clearly well understood and no real solution has yet been provided, although a number of experiments are underway. In addition to the examples given here, there is also a program in the U.K. to allow senior faculty to train in other disciplines ("discipline hopping" see below). It is much too early to tell which, if any, of these are successful in producing qualified researchers in systems biology. Given the importance of this issue and its embryonic state, some mechanisms for exchanging information internationally and locally on best practices are essential.

Selected Programs in Systems Biology

a. Graduate Programs with Systems Biology Courses

Europe and Great Britain

Flanders and Ghent University
Department of Plant Systems Biology
http://www.psb.ugent.be/
Max Planck Institutes
Institute of Molecular Genetics
Institute of Dynamics of Complex Systems

http://lectures.molgen.mpg.de/
http://www.mpi-magdeburg.mpg.de/

University of Rostock
Systems Biology & Bioinformatics Program
http://www.sbi.uni-rostock.de

University of Stuttgart
Systems Biology Group
http://www.sysbio.de/

Humboldt University Berlin
Institute for Theoretical Biology
http://itb.biologie.hu-berlin.de/

Department of Theoretical Biophysics
http://www.biologie.hu-berlin.de/~theorybp/

Free University of Amsterdam
BioMolecular Integration/Systems Biology
http://www.systembiology.net/topmaster/topmasterbmisbam.htm

University College, London
Centre for Mathematics and Physics in the Life Sciences
http://www.ucl.ac.uk/CoMPLEX/

University of Warwick
Interdisciplinary Program in Cellular Regulation
Molecular Organization and Assembly in Cells
http://www.maths.warwick.ac.uk/ipcr/

University of Oxford
Centre for Mathematical Biology
http://www.maths.ox.ac.uk/cmb

Asia

*A*Star Bioinformatics Institute, Singapore*
http://www.bii.a-star.edu.sg/

University of Tokyo
Graduate School of Information Science and Technology
http://www.i.u-tokyo.ac.jp/index-e.htm

Department of Computational Biology, Graduate School of Frontier Sciences
http://www.k.u-tokyo.ac.jp/renewal-e/course_jyoho/senkou-e.html

Kyoto University and University of Tokyo
Education and Research Organization for Genome Information Science
http://www.bic.kyoto-u.ac.jp/egis/

Keio University
Institute for Advanced Biosciences
http://www.iab.keio.ac.jp/

North America

Cornell, Sloan-Kettering, and Rockefeller Universities
Physiology, Biophysics & Systems Biology
Program in Comp. Biology and Medicine
http://www.cs.cornell.edu/grad/cbm/
http://biomedsci.cornell.edu

Massachusetts Institute of Technology
Computational and Systems Biology Initiative (CSBi), Biological Engineering Division
http://csbi.mit.edu/

Princeton University
Lewis-Sigler Institute for Integrative Genomics
http://www.genomics.princeton.edu

Stanford University
Medical Informatics (SMI) and BioX
http://smi-web.stanford.edu/

University of California Berkeley
Graduate Group in Computational and Genomic Biology
http://cb.berkeley.edu/

University of California San Diego
Department of Bioengineering
http://www-bioeng.ucsd.edu/

University of Toronto
Program in Proteomics and Bioinformatics
http://www.utoronto.ca/medicalgenetics/

University of Washington
Department of Bioengineering, Deptartment of Genome Sciences
http://www.gs.washington.edu/

Virginia Tech
Program in Genetics, Bioinformatics and Computational Biology
http://www.grads.vt.edu/gbcb/phd_gbcb.htm

Washington University
Computational Biology Program
http://www.ccb.wustl.edu/

b. Short Courses

Humboldt University
Berlin Graduate Program
Dynamics and Evolution of Cellular and Macromolecular Processes
http://www.biologie.hu-berlin.de/

Biocentrum Amsterdam
Molecular Systems Biology Course
http://www.science.uva.nl/biocentrum/

Cold Spring Harbor Laboratory
Course in Computational Genomics
http://meetings.cshl.org/

Institute of Systems Biology
Introduction to Systems Biology and Proteomics Informatics courses
http://www.systemsbiology.org

University of Oxford
Genomics, Proteomics and Beyond
http://www.conted.ox.ac.uk/cpd/biosciences/courses/short_courses/Genome_Analysis.asp

c. Emerging Initiatives

German Systems Biology Research Program
http://www.systembiologie.de/

Harvard University
Department of Systems Biology
http://sysbio.med.harvard.edu/

Manchester Interdisciplinary Biocentre (MIB)
http://www.mib.umist.ac.uk/

University of Texas Southwestern
Program in Molecular, Computational and Systems Biology
Integrative Biology Graduate Program
http://www.utsouthwestern.edu/utsw/home/education/integrativebiology/

NATIONAL PROGRAMS FOR SUPPORT OF SYSTEMS BIOLOGY

Systems biology is a relatively new discipline that involves the integration of engineering, physics, and biology. Its future depends on new sources of funding, since most existing funding programs have difficulty crossing disciplinary lines. Also, given that the discipline is at a relatively early stage in its development, the scientific activities represent basic research and will be primarily funded by the primary supporters of basic research, i.e. governmental entities. The following section provides an overview of some of the national programs supporting systems biology.

National Programs in the U.S.

The first efforts to directly support systems biology were a training program initiated in 1996 by a private foundation, the Burroughs-Welcome Fund, and, in 1998, a research support program begun by the National Institute of General Medical Sciences (NIGMS) of the National Institutes of Health (NIH). In more recent years, virtually every federal agency involved in supporting science has generated programs for the support of systems biology. (More details can be found in Cassman, 2005.)

Since the NIH is by far the largest supporter of research in the biological sciences in the U.S., it is not surprising that it contains the largest and most diverse array of programs supporting systems biology. These include programs for support of individual and institutional training, individual research grants, centers, and targeted disease-oriented studies, e.g. Integrative Cancer Biology (National Institute of General Medical Scien-ces, 2005; National Cancer Institute, 2005). In addition, new trans-institute programs have been established to develop "National Technology Centers for Networks and Pathways" and centers in "Metabolomics Technology Development" (National Institutes of Health, 2005). There are also significant programs at the National Science Foundation (NSF), such as "Quantitative Systems Biotechnology" and "Frontiers in Integrative Biol-ogical Research" (National Science Foundation, 2005); at the Department of Energy (DoE), through multi-institutional consortia focused on the analysis of microbial systems (Department of Energy, 2005); and at the Defense Advanced Research Projects Agency (DARPA), which has a large program to develop computational models and tools for *in silico* analysis (Defense Advanced Research Projects Agency, 2005).

These are programs that identify themselves as supporting quantitative approaches to biological networks. They leave out the much larger array of support mechanisms for proteomics and genomics, some of which include within them activities that are indistinguishable from the programs cited. Furthermore, there is even more support for research identified as investigator-initiated, i.e. not specifically promoted through identifiable programs.

National Programs Outside the U.S.

U.K.

The Engineering and Physical Sciences Research Council (EPSRC) Life Sciences Interphase (LSI) Programme "aims to fund high-quality research at the boundary between engineering and the physical sciences and the life sciences" (Engineering and Physical Sciences Research Council, 2005). These include networks "which are expected to lead to new collaborative multidisciplinary research proposals," and which funded for a total of £60,000 each for three years. There are also postdoctoral mobility awards and LSI doctoral training centers that encourage multidisciplinary training.

The Medical Research Council (MRC) has a program called "Discipline Hopping" run jointly with ESPRC and the Biotechnology and Biological Sciences Research Council (BBSRC) (Medical Research Council, 2005). As defined in the notice, "this scheme allows researchers who have a track record in their own field in the physical sciences to apply for funding to investigate and develop ideas, skills and collaborations in the areas of biological, clinical, health services and public health research. Alterna-tively, life science researchers can apply for funding to develop ideas, skills and collaborations with physical scientists." The awards are for three months to one year and are for no more than £60,000.

The BBSRC has also initiated a program for Centres for Integrative Systems Biology which will "integrate traditionally separate disciplines such as biology, chemistry, computer science, engineering, mathematics and physics in a programme of international quality research in quantitative and predictive systems biology" (Biotechnology and Biolo-gical Sciences Research Council, 2005). A number of new Centers will be awarded for up to £5 million each.

Germany

The Federal Ministry of Education and Research (BMBF) began discussions in 2001 to determine new funding strategies in bioscience, with a focus on cross-disciplinary activities. Their conclusions resulted in the promotion of systems biology in Germany through a network of centers of excellence. Individual research projects would be developed as collaborative projects between science and industry, and focus on the hepatocyte. It was planned to provide up to €50 million over five years. The initial funding began in January, 2004. Twenty-five groups are now supported with funding of €14 million over the next three years (Federal Ministry of Education and Research, 2005).

The support for systems biology also has contributions from independent research organizations. For example, the Helmholtz Association supports a number of research centers in biology and medicine, among which is the Heidelberg Cancer Research Institute (DKFZ), a key organization in the BMBF program. Additionally, it maintains a Network for Bioinformatics which provides coordinated access to bioinformatics resources. The Max Planck Institutes (MPI) are another key element in supporting systems biology. The MPI for Plant Physiology in Berlin; the MPI for Molecular Genetics in Potsdam; and the MPI for Dynamics of Complex Technical Systems in Magdeburg all support systems biology. Finally, there is significant activity in universities, such as the Theoretical Biology Department at Humboldt University, and through support from states within the federal republic. This is seen most notably in the state of Baden-Wuerttemberg life science centers, one of which is the Center for Biosystems Analysis in Freiburg.

Switzerland

The ETH Zürich, the University of Zürich, and the University of Basel have generated a collaborative project entitled "Systems*X*" intended to serve as a focus for systems biology in Switzerland (Systems*X*, 2005). The structure will accommodate collaborative efforts across disciplines and locations. To ensure integration between the several sites involved, the project management will be at the highest levels, comprising the president of the ETH Zürich, the rector of the University of Basel, the rector of the University of Zürich, the vice-president of research at the ETH Zürich, the vice-rector of research at the University of Basel, the pro-rector for research at the University of Zürich, research representatives from Novartis and Roche, plus the spokesperson for Systems*X*. Components of Systems*X* will include the Functional Genomics Centre Zürich, the Glycomics Initiative at ETH Zürich and the University of Zürich, the Oncology Cell Transfer Project at the University of Zürich, the Basel Bioinformatics Initiative, and the Life Sciences Training Facility at the University of Basel.

The Netherlands

Dutch initiatives in systems biology include the NWO (the Dutch equivalent of a hybrid between NSF and NIH) that is funding projects in bioinformatics from molecule to cell, and in computational biology, at a level of $6 million each. There are also initiatives in National Genomic Centers (funded at $200 million), including two that address some aspects of systems biology (Center for Medical Systems Biology at Leiden, Vrije Univ., Institute of Environmental and Energy Technology (TNO), and the Kluyver Center). Future efforts were described in the form of a set of focused program proposals that focus on organisms (*L. lactis* and *S. cerevisiae*) as well as tools (the Silicon Cell). Finally, there is an active center conducting systems biology in the Netherlands, at the Biozentrum Amsterdam.

Other European Activities

Active centers exist in Brussels (Free University), Sweden (University of Lund), and through a program that stimulates collaborations in science, including systems biology, between Sweden and Denmark. Additionally, a number of trans-European programs attempt to support activities across national boundaries. Many of these are supported through the European Commission, including an effort focused on computational biology. A recently funded program relevant to systems biology is a network of organizations involved in *in silico* simulation of biological systems with the goal of aiding in drug design. Coordinated by the Technical University of Denmark, the network will be comprised of 25 universities, a number of national medical agencies, and the pharmaceutical company Novo Nordisk. The European Union will provide €10.7 million over five years.

Japan

The primary mode of governmental support for the sciences is through MEXT (the Ministry of Education, Culture, Sports, Science, and Technology). The programmatic activities are then conducted by the Japan Science and Technology (JST) agency, whose mission is basic research and support of infrastructure. Another branch of MEXT is the Japan Soci-ety for the Promotion of Science (JSPS), which supports research, fellowships, and, particularly, international collaborations. In addition, several other agencies support science. For example, the Ministry of Economics, Trade, and Industry (METI) funds the Japan Biological Informatics Consortium (JBIC).

The Japanese government has initiated a number of very large-scale projects primarily in the areas of genomics and proteomics. This includes the Millenium Project, which incorporates national efforts in the rice genome, human genome diversity, and bioinformatics. Bioinformatics includes, among other things, structural and functional genomics, a number of databases, and the development of bioinformatics technologies. Other large-scale efforts include mouse and human full-length complementary deoxyribonucleic acid (cDNA) annotation programs and a high-throughput structural genomics effort.

Until recently there has been no national program directly targeted to systems biology. However, MEXT has initiated the Genome Network Program in 2004 that will include an investigation of the human genome network, most of which will be carried out by RIKEN, the Institute for Physical and Chemical Research (MEXT, 2005). Additionally there will be components which will integrate the information gained into a broader database of genomic and proteomic information; development of new genome analysis technologies; and spin-off applications to specific biological projects.

Although targeted efforts to systems biology are recent, other support has been available. For example, the JST database lists 16 funded applications under the rubric of systems biology (JST, 2005). Perhaps most striking is the number of well-appointed institutes that are doing systems biology in some form. The largest of these is RIKEN. It has five campuses in Japan plus several abroad. The site in Yokohama includes a Genome Science Center and a Plant Science Center, both of which have activities related to systems biology. Institutes have also been set up through funding of local prefectures. Examples are the Kazusa DNA Research Institute, largely supported by Chiba Prefecture, and the Institute for Advanced Biosciences, supported by Yamagata Prefecture and Tsuruoka City. Finally, industry is significantly involved in the academic and national/regional institutes, through a number of mechanisms. These include gift funding, collaborations, and commercial start-ups.

Conclusions

The U.S. remains one of the few countries that has a significant targeted investment in systems biology. A clear exception is Germany, which has developed a new initiative in the systems biology of hepatocytes, beginning in January 2004. National programs have also been initiated in Switzerland and the U.K. in the last few years, while the European Commission has acted as a catalyst for multi-national programs. Additionally, activities in systems biology are underway in many locations, as part of ongoing "traditional" governmental support programs. This is perhaps particularly noticeable in Japan. However, it is hard to avoid the conclusion that both the breadth and the scale of systems biology support from governmental entities are significantly greater in the U.S. than elsewhere in the world.

A possible caveat to this conclusion depends on the definition of systems biology. As noted in the Introduction, there is a distinction between "systems biology" and "systematic biology." Systematic biology, the high-throughput collection of targeted data sets, is a booming business everywhere, fuelled by the success of the genome project. Systems biology, the computational analysis of biological networks, is much more sparsely represented. Although this is also true in the U.S., encouragement of these activities through federal funding programs is significant and growing. It was slightly discouraging to see how frequently systems and systematic biology were conflated. Although data collection is clearly critical, it was not often the case that there was a connection between the data collected and its potential use in modeling and simulation of biological systems. In general, the future of systems biology worldwide depends on the support of programs which consider experimental and data-driven approaches together with the computational methods needed to model specific biological problems. Relatively few funding programs explicitly focus on this.

INFRASTRUCTURE

The term "infrastructure" is almost as ambiguous as "systems biology." It can mean anything from a new building to a simple laboratory spectrometer. The benefits to systems biology of buildings that house investigators with common interests may be of more significance than in other disciplines, given the requirements of interdisciplinary research, but it does not necessarily require any special attention in this report. Everyone wants new space, and the benefits may be real, but the arguments are too diverse and too tied to local circumstances for us to get involved. Similarly, laboratory instrumentation is an absolute requirement for any discipline that has an experimental base, but needs are too varied to provide arguments for any specific tool.

The infrastructure to be discussed in this section is limited to large-scale resources, specifically databases, software repositories, and centers. The term "large scale" is used to mean resources such as the Entrez databases which not only serve to centralize and index knowledge but provide a common core of data for many investigators and research areas. For example, the centralization, standardization and dissemination of sequence data at the National Center for Biotechnology Information (NCBI), the European Molecular Biology Laboratory (EMBL) and elsewhere have allowed the growth and improvement of algorithms for phylogeny, homology and other functional assignment which make up the core to nearly all molecular biology. This "network" effect has also been partly realized in the area of structural biology, but other data ranging from microbial phenotype to genetic manipulation data to functional genomics have not yet been similarly controlled, standardized and centralized to achieve the same degree of synergy (and thus quality control).

Many of these issues have been discussed at greater length in other chapters in this volume. The discussion here will focus on some specific concerns.

Software Repositories

The WTEC team's visits abroad as well as the team's knowledge of activities in the U.S. confirmed the existence of extensive activities in the development of software, which, especially in modeling and simulation, is a critical tool for systems biology. The WTEC panel saw little reason to believe that the state-of-the-art in this area is significantly more advanced in one country or area of the world rather than another. Indeed, it is frequently directed to similar goals. (One group at RIKEN even called their software YAGNS, "Yet Another Gene Network Simulator.") The reasons for this cottage industry in software are many, including the need to accommodate data derived locally; the requirement for visualization to accommodate specific requirements of collaborators; and a lack of knowledge of what is already available. In general, however, it is a terrible waste of time, money, and effort. At the moment, locally created software is practically inaccessible, even when the developers are willing to release it, since documentation is often so scanty that the barriers to use are prohibitive. (A more detailed discussion can be found in Chapter 4 in this volume.)

A reasonable set of expectations for software is that it should be interoperable, transparent to the user, and sufficiently well documented so that it can be modified and adjusted to circumstances. In systems biology there is the additional complication that the data sets used are frequently very diverse and often inconsistent with each other. For the developers and skilled users these problems may occasionally be overcome. However, the benefits of systems biology will only become manifest when working biologists, who are not themselves sufficiently trained to use such software, can manipulate and use these techniques. Admittedly, the translation of systems biology to a broadly based approach is complicated by the innumeracy of most biologists. Some modicum of mathematical training, which is now lacking, will be required, (see section on education). However, there is an immediate need to provide an opportunity for potential users to access and effectively use bioinformatics, modeling, and simulation software. One possible approach is to create a central orga-nization that would serve as a repository for systems biology software as well as serve as a mechanism for validating and documenting their utility and for standardizing the developers' interfaces and data input/output formats. Like central data repositories, having a central software repository with software engineering standards in place should create a network effect wherein synergy is created by the combined use and reconfiguration of tools for more sophisticated analysis.

Given that agencies worldwide are engaged in promoting the development of systems biology software (see items above describing support by Defense Advanced Research Projects Agency and the National Institutes of Health in the U.S.; the EC; and Japan), it would seem reasonable to create a structure that will preserve and enhance the benefits from these programs.

Databases

The problems with the diversity of software noted above are paralleled by the diversity in the way data used for modeling is collected, annotated, and stored. (These questions are discussed in more detail in Chapter 2) These issues are even more complex than for sequencing since systems biology is highly context dependent. In order for these data to be useful outside of the laboratory in which they were generated they must be standardized, presented using a uniform ontology, and annotated sufficiently so that the specific cell type, conditions of the medium, etc., are clearly reproducible. Systems biology often requires the use of multiple forms of data, e.g. metabolite and mRNA profiling, kinetic and thermodynamic measurements, etc. It is important to insure that the data is presented and annotated in a form that allows for all these data types to be effectively correlated. Additionally, one of the important functions of easy access to data and software is peer-review. In order to evaluate, in an ongoing way, the increasingly complex data, both raw and processed, the increasingly sophisticated analysis tools, and the increasingly less complete papers (that cannot include all information because of the very complexity of the experiments and tools), it is vitally important that reviewers and community have continuous access to the results and tools used to produce the literature. Dealing with this very complex issue will require a focused effort by the researchers involved as well as the funding agencies. It must be done soon.

Centers

The benefits of focused centers containing large numbers of scientists are almost offset by their shortcomings. Although there are synergies to be had from complementary groups of investigators, there is also the tendency of such structures to become sclerotic over time. They must be approached with caution. Nevertheless, there are examples of such centers being of great value. Perhaps the most clear-cut examples are those where economy of scale yields results not otherwise possible, such as the high-throughput sequencing projects. These can be found at a number of sites in the U.S. supported by the NIH for both DNA sequencing and protein structure determination; at RIKEN Yokohama, JBIRC, and Kazusa DNA Research Institute in Japan; at the Max Planck Institute of Molecular Plant Physiology and EMBL in Germany; and at the European Bioinformatics Institute (EBI) in the U.K.

A different kind of center is one created around specific biological problems. It is possible to identify several examples of these related to systems biology. In the U.S. there are five centers established by the National Institute of General Medical Sciences of the NIH as well as the private Institute for Systems Biology in Seattle. Additionally, a number of centers are being planned through the NIH trans-agency "Roadmap" programs. One recent award, part of the National Centers for Biomedical Computing, was to Stanford University for "a simulation toolkit that enables scientists worldwide to model and simulate biological systems from single atoms to entire organisms." The Department of Energy has created centers focusing on bacterial systems. Another interesting approach is that of the Alliance for Cell Signaling which plans a comprehensive analysis of the signaling molecules and pathways in eukaryotic cells. In contrast to other centers, which are co-located, this group involves seven laboratories in five different locations. This was supported by a mechanism called "glue grants," offered by the National Institute of General Medical Sciences of the NIH, and was specifically designed to coordinate focused research efforts across multiple institutions.

In Germany, the Max Planck Institutes provide an institutional framework that makes it possible to create centers focused on specific problem areas. An example is the Institute for Complex Technical Systems in Magdeburg, Germany, headed by Prof. Ernst Dieter Gilles. This is an institute with a number of related focus areas in engineering, one of which is in systems theory. This group has 17 faculty members, all of whom work in some aspect of systems biology. Although they have some capability in experimental areas, their strength is in the depth and breadth of theory, and they develop this through a wide array of external collaborations in experimental programs. Another example is the Max Planck Institute for Plant Physiology in Golm, Germany, headed by Professors Lothar Willmitzer and Mark Stitt. The departments in this Institute are effectively merged, with the goal of an integrated research approach to solve basic questions in plant metabolism, combining methods from genetics, molecular biology, and chemistry. The strengths of such an integrated approach were particularly visible in the linkage of the bioinformatics capability of the Institute to all of its laboratories, and in the existence of an overall strategy for addressing phenotyping coupled to the disruption of critical genes, all linked to a comprehensive sequencing and bioinformatics infrastructure. Most recently, a distributed program focused on the systems biology of the hepatocyte has been established. This is most comparable to the Alliance for Cell Signaling in the U.S.

In the Netherlands, Prof. Hans Westerhoff at the Free University in Amsterdam has a group combining experimental and theoretical appr-oaches to systems biology, much of it oriented around metabolic control analysis and hierarchical control analysis. The existence of a group of investigators working in related areas has a clear synergistic influence, not only in research but in training as well.

In Japan the government, both centrally and at local levels, has created a number of institutes targeting specific problem areas. Most of these are in the area of large-scale data generation, organized collections of full-length cDNA clones, and in collection of mutant lines. However, several are more closely linked to systems biology. As examples, it is worth mentioning two that are part of Keio University and are largely the creation of strong leaders. Prof. Hiraoki Kitano has developed the Symbiotic Systems Project in Tokyo, while Prof. Masaru Tomita has initiated the Institute for Advanced Biosciences in Tsuruoka. Both of these investigators have a long history of research in systems biology and have created programs which link theory and experiment. In addition, the National Institute for Advanced Science and Technology (AIST) has created a novel research structure, the Computational Biology Research Center (CBRC) in Tokyo. The CBRC is organized to conduct research in bioinformatics only in terms of information theory completely independent from experimental biology projects, although they engage in many collaborations. Areas of emphasis include genome informatics, molecular modeling and design, and cellular informatics.

A variety of centers are being created in the U.S., largely to provide technology development. An argument could be made for the creation of centers targeted to specific research problems, and specific experimental systems, in systems biology in the U.S. The need for consistent and reproducible data and the need for close collaboration between theorists and experimentalists are both arguments for co-located groups that can interact easily and often. It is also far easier to enforce standards at such centers. There are a number of examples of such organizational structures outside the U.S. In general, they are characterized by strong leaders with clear programmatic goals. These kinds of centers are relatively rare in the U.S., at least in part because the scientific culture is oriented to smaller groups with more distributed authority. This has in fact been a major strength of the U.S. system for many years and in most research fields. However, it should not blind us to the possibility that other approaches can supplement this model. At this point in time, systems biology can benefit from stronger centralized approaches that will allow the testing of model systems in an optimum environment.

Conclusions

The key issues of infrastructure in systems biology—databases, software, and centers—are common across all the countries involved. In particular, the availability of a common structure for the use of data and software is lacking and requires immediate attention. International collaborations will be needed to accomplish this.

REFERENCES

Biotechnology and Biomedical Sciences Research Council, http://bbsrc.mondosearch.com/ (Accessed October 1, 2005)

Cassman, M., (2004). Systems Biology in the U.S. *IEE Sys Biol* 1: 204–205.

Defense Advanced Research Project Agency, http://www.darpa.mil/ipto/programs/biocomp/index.htm (Accessed October 1, 2005)

Department of Energy, http://doegenomestolife.org/ (Accessed October 1, 2005)

Engineering and Physical Sciences Research Council, http://www.epsrc.ac.uk/ ResearchFunding/Programmes/LifeSciencesInterface/default.htm (Accessed October 1, 2005)

Federal Ministry of Education and Research, http://www.bmbf.de/en/1140.php (Accessed October 1, 2005)

JST, http://j-east.tokyo.jst.go.jp/cgi-bin/DbSearchList.cgi, (Accessed October 1, 2005)
Medical Research Council, http://www.mrc.ac.uk/index/funding/funding-specific_schemes/funding-current_grant_schemes/funding-discipline_hoppers.htm (Accessed October 1, 2005)
MEXT, http://www.mext-life.jp/genome/english/ (Accessed October 1, 2005)
National Cancer Institute, http://dcb.nci.nih.gov/newsdetail.cfm?ID=11 (Accessed October 1, 2005)
National Institute of General Medical Sciences, http://www.nigms.nih.gov/funding/ complex_systems.html (Accessed October 1, 2005)
National Institutes of Health, http://nihroadmap.nih.gov/ (Accessed October 1, 2005)
National Science Foundation, http://www.nsf.gov/funding/pgm_summ.jsp?pims_id=6188&org=EF (Accessed October 1, 2005)
Systems*X*, http://www.bsse.ethz.ch/ (Accessed October 1, 2005)

APPENDIX A: PANELIST BIOGRAPHIES

MARVIN CASSMAN (CHAIR)

A native of Chicago, Marvin Cassman received his BS and MS degrees from the University of Chicago. He earned a PhD in biochemistry in 1965 at the Albert Einstein College of Medicine of Yeshiva in New York. Following a postdoctoral fellowship at University of California (UC) Berkeley, he joined the UC Santa Barbara faculty, leaving 1975 to follow an appointment to the National Institutes of Health (NIH).

Dr. Cassman was the director of the National Institute of General Medical Sciences at the National Institutes of Health (NIH) from 1994–2002 and was regarded as a pioneer in perceiving new trends in biomedical science and creating new approaches to meet ideas of opportunity or need.

He left NIH in May 2002 to become the executive director of the California Institute for Quantitative Biomedical Research (QB3). QB3 is a consortium of three University of California schools—San Francisco, Berkeley, and Santa Cruz. The intent of the Institute is to integrate the physical, mathematical, and engineering sciences with biomedical research.

Dr. Cassman left QB3 and the University of California in 2004. Dr. Cassman currently serves on several advisory committees and his honors and awards include the 1991 Presidential Meritorious Executive Rank Award, the 1983 NIH Director's award, and The Biophysical Society Distinguished Service Award in 2003.

ADAM ARKIN

Adam Arkin received his BS degree from Carlton College in 1988 and his PhD in physical chemistry from MIT in 1992. He was a postdoctoral fellow at Stanford University from 1992–1997. He has been a faculty scientist of the Lawrence Berkeley National Laboratory since 1998, assistant professor of bioengineering and chemistry at the UC Berkeley since 1999, and investigator of the Howard Hughes Medical Institute since 2000.

His research is in systems and synthetic biology using quantitative measurement and modeling of regulatory network dynamics. His lab employs tools from physical chemistry, nonlinear dynamics and control theory, molecular biology and computational biology to explore issues in bacterial and mammalian signal transduction and viral development.

Dr. Arkin received the 1999 MIT Technology Review Top 100 Most Innovative Young Scientists Award.

FRANK DOYLE

Frank Doyle holds the Duncan and Suzanne Mellichamp Chair in Process Control in the Department of Chemical Engineering at the University of California at Santa Barbara (UCSB), as well as appointments in the Electrical Engineering Department, and the Biomolecular Science and Engineering Program. At UCSB, he is the associate director of the $50 million Institute for Collaborative Biotechnology funded by the Army Research Office.

He received his BSE from Princeton (1985), his CPGS from Cambridge (1986), and his PhD from Caltech (1991), all in chemical engineering. Prior to his appointment at UCSB, he held faculty appointments at Purdue University and the University of Delaware, and held visiting positions at DuPont, Weyerhaeuser, and Stuttgart University.

He is the recipient of several research awards (including the NSF National Young Investigator, ONR Young Investigator, and Humboldt Research Fellowship) as well as teaching awards (including the Purdue Potter Award, and the ASEE Ray Fahien Award). He is currently the editor-in-chief of the IEEE Transactions on Control Systems Technology, and holds associate editor positions with the Journal of Process Control, the SIAM Journal on Applied Dynamical Systems, and Interface. His research interests are in systems biology, drug delivery for diabetes, and control of particulate processes.

FUMIAKI KATAGIRI

Fumiaki Katagiri received his BS and MS degrees in chemistry from Kyoto University. In 1991 he earned his PhD in life sciences from Rockefeller University in New York. He did postdoctoral work in genetics at Harvard Medical School from 1991–1995.

Dr. Katagiri was assistant professor in the Department of Biological Sciences of the University of Maryland, Baltimore County, from 1995–1998 and senior staff scientist of the Torrey Mesa Research Institute from 1998–2003. In 2003 he was appointed associate professor in the Department of Plant Biology at the University of Minnesota.

Dr. Katagiri is widely published and the author or co-author of seven publications since 2003. His main research interests focus on issues in systems biology of plant disease resistance.

DOUGLAS LAUFFENBURGER

Douglas Lauffenburger is an Uncas & Helen Whitaker professor of bioengineering in the Biological Engineering (BE) Division, Biology Department, and Chemical Engineering Department and is a Member of the

Center for Cancer Research, Center for Biomedical Engineering, and Biotechnology Process Engineering Center at MIT. He serves as director of the BE Division, and as chair of the Executive Committee of the MIT Computational Systems Biology Initiative.

Dr. Lauffenburger's BS and PhD degrees are in chemical engineering from the University of Illinois and the University of Minnesota, in 1975 and 1979, respectively.

His major research interests are in cell engineering: the fusion of engineering with molecular cell biology. A central focus of his research program is in receptor-mediated cell communication and intracellular signal transduction, with emphasis on development of predictive computational models derived from quantitative experimental studies, for cell cue/signal/response relationships important in pathophysiology with application to drug discovery and development.

He is a member of the National Academy of Engineering and of the American Academy of Arts & Sciences, and has served as president of the Biomedical Engineering Society, chair of the College of Fellows of AIMBE, and on the Advisory Council for the National Institute for General Medical Science at NIH.

CYNTHIA STOKES

Cynthia Stokes received her BS in chemical engineering from Michigan State University and her PhD in chemical engineering from the University of Pennsylvania. She did postdoctoral work at the National Institutes of Health.

She is principal scientist, Immunologic Diseases, *In Silico* R&D at Entelos, Inc. Her research in systems biology at Entelos uses mathematical modeling of multi-scale, dynamical, biological systems for pharmaceutical discovery and development. She is involved in the development of large-scale mathematical models of immunologic diseases, including asthma and rheumatoid arthritis, and the use of these models for basic and applied research.

Applications include basic research on complex biological systems and mechanisms of disease dynamics, drug target identification, prioritization and validation; chemical lead evaluation, and clinical trial design.

Prior to joining Entelos, Dr. Stokes was assistant professor at the University of Houston in the Chemical Engineering Department.

OTHER TRAVELING TEAM MEMBERS

Semahat Demir

Semahat Demir received her BS degree in electronics engineering from Istanbul Technical University (1988), MS degree in biomedical engineering from Bosphorous University (1996), and second MS degree (1992) and PhD degree (1995) in electrical and computer engineering from Rice University. She did her postdoctoral training in the biomedical engineering department at Johns Hopkins University (1995-96). In industry, Dr. Demir worked as the technical manager and medical laser engineer for Messerschmidt Bolkow Blohm and Rodenstock, in Turkey (1988-1989). She was also a research and development engineer in the X-ray Division of the Medical Engineering Center of Siemens Company in Erlangen, Germany (1988). Dr. Demir has been program director, Biomedical Engineering and Research to Aid Persons with Disabilities, Division of Bioengineering and Environmental Systems, at National Science Foundation since June 1, 2004 while she is on a leave as a professor of biomedical engineering from the Joint Biomedical Engineering Program of University of Memphis and University of Tennessee Health Science Center. Dr. Demir's current research is in the area of computational bioengineering. Her research program integrates research, education and training, and emphasizes mathem-atical modeling and computer simulations in both cardiac electrop-hysiology and neuroscience.

Fred Heineken

Fred received his BS degree in chemical engineering from Northwestern University in 1962 and his PhD in chemical engineering from the University of Minnesota in 1966. His thesis was on the "Mathematical Aspects of Enzyme Kinetics." Following graduate school, he worked for Monsanto for five years where he did enzyme product development work. Heineken then joined the University of Colorado where he did research in respiration physiology and teaching of chemical engineering. After four years at the University of Colorado, Heineken joined COBE Laboratories where he worked on hemodialysis research and product development. After nine years at COBE, he joined the National Science Foundation (NSF) as a program director doing funding of biotechnology and biochemical engineering in the Engineering Directorate. Heineken continues his work as a program director in the Engineering Directorate and recently received an NSF Award for Meritorious Service.

WTEC Team Members

Other traveling members include: Roan Horning (WTEC), Hassan Ali (WTEC), and Gerald Hane (Q-Paradigm).

APPENDIX B: SITE REPORTS—EUROPE

Site: **Cambridge University**
Dept. of Anatomy
Downing St., Cambridge, CB2 3DY, U.K.

Date visited: July 9, 2004

WTEC Attendees: F. Katagiri (Report author), D. Lauffenburger, C. Stokes

Hosts: Dennis Bray, Tel: (+44) 1223 333 771, Email: Db10009@ca.ac.uk
Matthew Levin (post-doc), Karen Lipkow (post-doc), David Odde (professor on sabbatical from the University of Minnesota)

BACKGROUND

Bray started modeling *E. coli* chemotaxis in the mid 1990s after he did a sabbatical at Cal Tech. *E. coli* has the simplest chemotaxis regulation known—only six or seven proteins are involved. In addition, the chemotaxis regulation network is fairly well isolated from other systems. Bray's approach to the network is to build as detailed a computer model as possible by incorporating all the trustworthy data. His major collaborator in experimental work is Bob Bourret at the University of North Carolina, Chapel Hill.

CURRENT PROJECTS

1. Deterministic ordinary differential equation (ODE) model (Levin)
2. Stochastic model of the receptor complex
3. Stochastic 3D diffusion model (Lipkow)

LEVIN'S PRESENTATION

BCT (bacteria chemotaxis) model—deterministic ODE model.

Network components: Tar receptor (aspartate receptor), CheR (receptor methylase), CheB (receptor demethylase), CheW, CheA (CheY kinase), CheZ (CheY phosphatase), CheY, FliM (which determines which way the motor rotates: when phosphorylated CheY binds FliM, the motor rotates clockwise, which leads to tumbling).

The model can simulate excitation (binding of an attractant to a receptor leads to inhibition of CheA; consequently, the motor keeps rotating counter-clockwise, which is the run state), adaptation (extended stimulation by an attractant leads to receptor methylation, which desensitizes the receptor; this state is reversible under an extended no-attractant condition), mutant phenotype, and individuality. In pyBCT (written in Python), the simulated swimming behavior is visualized. A bacterium swims in a 2D space with a 1D aspartate gradient.

Lauffenburger raised the point that individuality can arise from stochastic events or heterogeneity of parameters (e.g., molecule concentrations) in each individual.

LIPKOW'S PRESENTATION

Smoldyn model—Smoluchowski dynamics-based stochastic simulator.

The states and positions of molecules, such as CheY, are visualized through time in a confined 3D space that represents a bacterium. A newly developed Brownian dynamics program is used for calculation. The model is being used to investigate how the architecture of the intracellular space affects signaling and ultimately chemotaxis. Simulations were shown for investigations of the importance of cell size, molecular positioning (e.g., CheZ in a polar cluster versus freely diffusing in the cytoplasm), varying diffusion constants, macromolecular crowding, and motor response.

SYSTEMS BIOLOGY COMMUNITY AT CAMBRIDGE

Neither formal organization nor established informal organization exists. The Institute of Mathematics invites seminar speakers from physics and biology.

EDUCATION AT CAMBRIDGE

No formal undergraduate or graduate program exists.

Site: **European Bioinformatics Institute (EBI)**
Wellcome Trust Genome Campus, Hinxton
Cambridge, CB10 1SD, U.K.

Date visited: July 9, 2004

WTEC Attendees: D. Lauffenburger (Report author), C. Stokes, F. Katagiri

Hosts: Nicolas Le Novere, Group Leader, Computational Systems Neurobiology program, Tel: (+44) 1223 494 521, Email: lenov@ebi.ac.uk

BACKGROUND

Central Theme

- Models of neuronal cell biology—NOTE: differentiates neurobiology from neuroscience; the former includes molecular/cell mechanisms whereas the latter emphasizes abstract network behavior
    1. Signaling pathway topologies
    2. Dynamics of signaling pathways
        a. Ordinary differential equation (ODE) models
    3. Spatio-temporal organization of synapse components
        a. Compartmental ODE models

Context

- EBI was originally a data resource, similar to the National Center for Biotechnology Information (NCBI), but more recently has taken on a primary research role; multiple research groups, most of which are informatics-centric
    1. Functional genomics
    2. Structural genomics
    3. Text mining
    4. Network inference
    5. Imaging
    6. Computational neurobiology
- There is a strong database emphasis, but they are moving from genome sequence to expression to networks

SPECIFIC PROJECTS

- DopaNet—European collaborative consortium aimed at modeling mesotelencephalic dopamine signaling
    1. Initial funding from European Science Foundation for organization meetings
    2. Hope to gain research program funding from other EU agencies
- Stages
    - 1: ontology/database development; literature mining for quantifiable properties in this network; establishment of tools for modeling and simulation
    - 2–3: literature mining for other components; generation of dynamic data; modeling and simulation
- Experimental collaborators

    Other European efforts in Systems Neurobiology and realistic modeling of neurons

    1. Seth Grant—Wellcome-Trust Sanger Institute
    2. Nigel Goddard—Edinburgh
    3. Erick DeSchutter—Antwerp
    4. Rolf Kotter—Düsseldorf

PERSPECTIVES

- Large-scale databases will be useful foundations generally, but will not deal with more specific needs of specialized research programs
    1. Local, smaller-scale databases will be needed on top of the large-scale, broad-community efforts
        - Tools for efficient local development are important needs
- Has been conflict between post-genomics community and modeling community regarding the value of 'systems biology,' but the post-genomics community has very recently been attempting to take on this mantle

Site: **European Commission (EC)**
European Commission Office
8 Square de Meeus B-1049 Brussels, Belgium

Date visited: July 8, 2004

WTEC Attendees: F. Heineken (Report author), M. Cassman, A. Arkin, F. Doyle, H. Ali

Hosts: Frederick B. Marcus, Tel: +32-2 29-58337,
Email: Frederick.marcus@cec.eu.int,
Philippe Jehenson, Tel: +32-2-298-6454,
Email: philippe.jehenson@cec.eu.int
Olaf Wolkenhauer, University of Rostock,
Tel:/Fax: +49(0)381-498-33-35/36,
Email: wolkenhauer@informatik.uni-rostok.de

BACKGROUND

Frederick Marcus has training as a physicist and attended MIT and Oxford University. Before joining the European Commission (EC), he had research experience in nuclear fusion. While at the EC, Marcus has participated in formulating intellectual property rights policies for the Sixth Framework. He is currently principal scientific officer for the Fundamental Genomics and Bioinformatics Unit (Unit F4) at the EC, and has a budget portfolio of approximately €50 million for projects lasting three to five years.

Phillippe Jehenson is principal scientific officer for the Biotechnology and Applied Genomics Unit at the EC.

Olaf Wolkenhauer obtained training as a control engineer from the University of Hamburg and additional training in biology (PhD) from the University of Portsmouth in the United Kingdom. He currently holds the first chair in systems biology in Germany, is a professor in computer science at the University of Rostock, and is a recent awardee of a Specific Targeted Research Projects (STREPS) project in computational systems biology. Olaf is also editor of the new journal "Systems Biology."

PERTINENT ACTIVITIES OF THE EUROPEAN COMMISSION

Fred Marcus led the discussion and provided the group a handout with an agenda for the meeting.

The discussion of the group centered on the activities associated with the Sixth Framework, which covers the time frame 2002–2006.

The Research and Technological Development (RTD) Budget for the EC is about €17.5 billion for 2002–2006, which is about 4–5% of the total RTD Budgets of the member states. The EC activities focus on arranging collaborations amongst the member states. Projects require the participation of at least three member states. The philosophy behind this is to encourage better communication amongst the member states. RTD projects from one country are evaluated and funded directly by that country. Industrial organizations can apply for EC RTD funds.

Some countries who are not EC Members (i.e., Switzerland, Israel, Norway, Iceland, Turkey, etc.) have joined as associate members for purposes of participating in the EC RTD activities. These associate members are expected to contribute funds to the EC for RTD activities.

A number of "New Instruments" exist for funding projects in the "European Research Area" (ERA). These include:

- Networks of Excellence
- Integrated Projects
- Specific Targeted Research Projects
- Specific Support Activities (SSA)
- Coordination Action (CA)

Networks of Excellence are typically five-year projects with total budgets of €8–12 million. These projects can be extended for a total duration of up to seven years, and are typically not renewable.

Integrated Projects have similar size and budgets, but are more results oriented, whereas Networks of Excellence are more networking oriented. Of necessity, both have management and goal orientations, but they are distinguished by different emphasis. STREPS are three-year projects at a funding level of €1.5 to 2.5 million per year.

Frederick Marcus has a strong interest in bioinformatics, computational systems biology (CSB) and systems biology (SB) where the emphasis is on computational methods for describing biological systems. This is part of Sixth Framework, and there are calls for proposals in these areas. It is planned to have four calls for proposals as a means of implementing the Sixth Framework.

Systems biology will be a key area in the Seventh Framework, which is scheduled to start early in 2007. Topical areas like SB and CSB need to have the agreement of the member states of the EC, and negotiations amongst the EC member states for detailed topical areas for Seventh Framework are currently underway.

Two reports have been issued, for a CSB Workshop held in Brussels on September 10–11, 2003, and for an EU Projects Workshop on Systems

Biology on December 8, 2004. Eight "Areas for Action" were identified and discussed briefly in the team's meeting with Frederick Marcus. More information on these workshops and other topics is available at: http://www.cordis.lu/lifescihealth/genomics/home.htm.

RESEARCH PROJECTS

A wide variety of research projects are financed by the EU. These projects involve a directly funded and strong research component as well as co-ordination, which additionally mobilizes major resources of the participating laboratories. The largest of these projects create distributed virtual institutes, e.g. BIOSAPIENS, with capabilities, note American reviewers, that exceed those of comparable U.S. projects.

Name	Topic (Number of Participating Laboratories) Website
ATD	Alternate transcripts (9) http://www.atdproject.org
BIOSIM	Cellular, physiological, pharmacological SB (40) http://chaos.fys.dtu.dk/biosim
BIOSAPIENS	Genome Annotation, Bioinformatics School (24) http://www.biosapiens.info
COMBIO	P53-MDM2 Spindle SB (8) http://www.pdg.cnb.uam.es/COMBIO
COSBICS	Signaling SB (6) http://www.sbi.uni-rostock.de/files/cosbics_dec_04.pdf
DIAMONDS	Cell cycle, yeast-human SB (10) http://www.sbcellcycle.org/main.htm
EMBRACE	Bioinformatics Grid (17) http://www.embracegrid.info
EMI-CD	SB models-complex disease (5) http://pybios.molgen.mpg.de/EMICD/
ENFIN	Bioinformatics basis for SB modeling (17) http://www.ebi.ac.uk
ESBIC-D	SB of cancer-patients (7) http://www.molgen.mpg.de/research/lehrach/
EUSYSBIO	Coordination of SB (10) http://www.eusysbio.org/
YSBN	SB of yeast cells (18) http://www.gmm.gu.se/YSBN/
TEMBLOR	BioinformaticsIntegration, ArrayExpress, ProteinProteinInteractions, protein structure (25) http://www.ebi.ac.uk/Information/funding/temblor.html

These projects may be grouped as follows:

Bioinformatics: ATD, BIOSAPIENS, EMBRACE, TEMBLOR. In the largest of these, starting at the beginning of 2002, the EC provided the European Bioinformatics Institute (EBI) in Hinxton, U.K., with €20 million for three-year duration to support TEMBLOR, a project to allow users to carry out complex inquiries across databases. This project involves 25 collaborators in 12 countries, with EBI taking the lead. BIOSAPIENS and EMBRACE are Networks of Excellence which use multimillion euro yearly funding to produce a Virtual Institute of Genome Annotation, a European School for Bioinformatics, and a Bioinformatics Grid to seamlessly link the major databases in Europe. ATD is a smaller research project on alternate transcripts

Systems Biology Research: The large Networks of Excellence combine research and integration, with ENFIN providing database and tool assembly for systems biology, and BIOSIM developing *in silico* modeling for cellular, physiological and pharmacological SB. The medium-size SB research projects COMBIO, COSBICS, DIAMONDS, EMI-CD use SB to explore different processes at the cellular level.

Systems Biology Coordination: There is an SSA for the coordination of systems biology in Europe called EUSYSBIO. It is ongoing (2003–2005) and involves eight European partners, which are coordinated by the Research Center in Julich, Germany. There are also two coordination actions to assemble SB tools and coordinate research; ESBIC-D for complex diseases, concentrating on cancer and unifying computation, lab research and clinical data from patients; and YSBN for yeast research.

Site: **German Cancer Research Center (DKFZ)**
Deutsche Krebsforschungszentrum
Im Neuenheimer Feld 280, 69120 Heidelberg, Germany

Date Visited: July 7, 2004

WTEC Attendees: A. Arkin (Report author), M. Cassman, F. Doyle, F. Heineken

Hosts: Otmar Wiestler, Professor, CEO and Chief Science Officer, DKFZ,
Dr. Siegfried Neumann, Professor, Office of Technology, Merck KGaA,
Tel: +49-6151-722 151,
Email: siegfried.Neumann@merck.de,
Ursula Klingmüller, Professor, DKFZ,
Tel: +49-6221-42-4481,
Email: u.klingmueller@dkfz-heidelberg.de,
Prof. Dr. Willi Jäger,
Prof. Dr. Wolfgang Driever,
Matthias Reuss,
Dr. Eric Karsenti,
Dr. Jens Doutheil,
Prof. Dr. Jan Hengsteler,
Sven Sahle

BACKGROUND

Dr. Ursula Klingmüller made general welcomes and Dr. Otmar Wiestler made the formal introduction to the Institute, outlining the research areas at the Deutsche Krebsforschungszentrum (DKFZ) and its contributions in the Hepatocyte Alliance.

DKFZ Heidelberg is a non-profit, research center with a focus on cancer research. It was founded in 1964 and since 1975 it has been a national research center with funding from both state and federal agencies. The Institute is a member of the Helmholtz association, which is a public research association that funds concerted research in a number of areas. There are 15 institutions in Germany under its umbrella. The total staff of the DFKZ exceeds 2,000 with research topics spanning: signaling pathways, cell biology and cancer, structure and functional genomics, cancer risk factors and prevention, tumor immunology, innovative diagnostics and therapy, infection and cancer, and cancer stem cells. He stressed that the current focus is to increase research efforts in cancer stem cells, pharmacogenomics, molecular diagnostics and systems biology. The resources and coordination seemed equal to the task.

The turnout for the panel in the present meeting was excellent and included scientists from many regional research centers and program directors from funding agencies.

RESEARCH PROJECTS

The WTEC team then heard briefly from Prof. Dr. Siegfried Neumann, from the Office of Technology at Merck KGaA and a member of the Steering Committee for the Federal Ministry of Education and Research (BMBF) Funding Initiative "Systems for Life-Systems Biology," which was the main driver for the formation of the Hepatocyte Alliance. We heard briefly about the rationale for choosing a cell as difficult as the hepatocyte for the Alliance and about the industry perspective on systems biology. He believes that systems biology currently is some distance from effective and efficient praxis in industry.

Later in the day, Neumann (with help from Drs. Gisela Miczka, and Roland Eils) further described the Hepatocyte Alliance and how it came to be. In 2001 the BMBF initiated a program in systems biology which was to fund medium- to long-term research programs. A synergy existed between this program and with existing BMBF initiatives in genomics and proteomics. To evaluate proposals and to insure they meet the international state-of-the-art, a core expert panel of nine scientists and an external panel of more than 70 were assembled to discuss, in four workshops, the structuring, thematic focus priority of such a program and to provide recommendations. Ultimately the goal to understanding the mechanisms of behavior of hepatocytes was settled on because of: 1) their central function in metabolism, 2) their central role in the uptake and conversion of drugs, and 3) their ability to regenerate. They expected this research to have a substantial impact on problems in pharmacology and pathophysiology. They were (and are) fully aware of the challenges involved in the high complexity of the cells, the difficulty in their handling and cultivation, and in the creation of a bioinformatic and modeling framework to organize information about these cells. However, such challenges seem appropriate to a long-term, grand challenge project.

The approach of BMBF in setting up this Alliance was to create an interdisciplinary competence network linking bioscience with computer science, mathematics, physics and engineering sciences. They started with studies on defined biological functions of the cell with establishing standardized primary cells, methods and tools. Two sub-projects rest on two methodological and cell biological platforms. Project A is detoxification and dedifferentiation in hepatocytes and Project B is regeneration of hepatocytes. The two platforms are cell biology and modeling.

The project has 25 total participating groups. Funding of €14 million has been provided over three years beginning in 2004. There is an external steering committee composed of: Dieter Osterhelt (MPI Biochemistry, Munich), Roland Eils (DKFZ Heidelberg), Joseph Heijnen (Technical Uni. Delft.), Karl Kuchler (Institute for Medical Biochemistry, University of Wien), Siegfried Neumann (Merck KGaA), and Hans Westerhoff (Biocentrum Amsterdam).

Neumann was followed by a presentation by Prof. Dr. Willi Jäger from the Interdisziplinäres Zentrum für Wissenschaftliches Rechnen (IWR, the Interdisciplinary Center for Scientific Computing) at the University of Heidelberg. He discussed a new large initiative called BioMS (Modeling and simulation in the biosciences) in the context of the long history of mathematical modeling and its applications to biology, particularly citing the University of Minnesota, Minneapolis, for having spawned many of the early pioneers. The historical difficulty in creating such programs, he says, has largely been due to the perception that models at the cellular level of biology had little application and certainly weren't ready for industrial applications. However, with the BioMS program, a center focused at this level is coming together with current funding for four postdocs and three junior professors ($250,000/year to do independent work). They are also getting a building for this effort out of the Bioquant program. He cites that such a building provides the necessary localization both for intellectual critical mass and for the administration to take the unit seriously. (This last is, in my opinion, an especially important and novel point.) Bio MS is a multi-institutional structure established to promote the use of modeling and simulation in the biosciences. It is funded by several sources, i. e. the state of Baden-Wuerttemberg, a private foundation, and by participating institutions, one of which is the University of Heidelberg.

The BioQuant program in general is providing a great deal of support over the next four years. The original program was to initiate and support cooperative projects of groups in biological and biomedical basic research and of mathematical and computational sciences. There was an internal competition which received 60 research topics descriptions many calling for for mathematical modeling help. The selected topics for this center are: mechanism of transport and reactions in membranes; molecular modeling and simulation of dynamics and conformational change of complex molecules; modeling and simulation of biochemical, biophysical processes for biogenesis and activities of vesicles; modeling of signal transduction; modeling neural networks; molecular biomechanics of cells; interaction biochemistry and biomechanics.

The BioQuant center was approved in January 2002 by the state of Baden-Wuerttemberg. The initial expenses were €23.31 million for the building, €3.57 million for basic equipment. The building is a total of 5100 m^2 with 2439 m^2 going to life sciences and 1090 m^2 going to modeling. The building is located in the center of the campus in Heidelberg, near physics and mathematics and not too far from the DKFZ. It is designed to house young researchers rather than more senior professors; it is an incubator of thoughts. They are also initiating training programs in biotechnology and attempting to review the biology curriculum with more mathematics both at the college and gymnasium level.

Prof. Dr. Wolfgang Driever, a developmental biologist from the faculty of biology at the Freiburger Zentrum für Biosystemanalyse, University of Freiburg, spoke on the requirements for success in systems biology. He stressed the needs for standard systems for the acquisition of quantitative time-resolved data, for the analysis and modeling of that data, as well as for the need for quality control efforts and assurance. The work at Freiburg is established within the framework of the state of Baden-Wuerttemberg life science centers and is geared towards integrative research in genomics, proteomics, metabolomics and imaging. Project areas mentioned included cellular systems regulation with a focus on single-cell systems as independent regulatory units, and analysis of complex regulatory circuitries. He stressed the value of using a few standard cell systems to gain a "network effect" from many researchers working on just a few systems (such as the hepatocyte); Analysis of regulatory networks that control supracellular organization including elucidating the complexity of interactions which control differentiation, complex multicellular systems, patterns in brain development, etc. There are also expanding studies in metabolomics, mainly in plants, with the aim of establishing models for metabolite flow by profiling low molecular weight (MW) metabolites and secondary metabolites and understanding the kinetics and transport of these molecules. One example is vitamin A synthesis in golden rice. Starting with Eric Davidson's example of developmental modeling, Driever wishes to extend such approaches into application to signal transduction pathways. He cites the work by Ursula Klingmüller, discussed later, as an example of successful application, successful in part, because of the tight coupling between theorists such as Jens Timmer at Freiburg University and biologists such as Klingmüller.

The efforts at Freiburg also have their own building and many small groups are starting projects within it. They have successfully centralized data acquisition and modeling together. Importantly, they are a highly interactive group with outreach to 15 related faculty members in the computer science department and with the 18 faculty and 200 researchers at the new Freiburg Institute for MicroSystems Technology. There is also an attempt to initiate a tri-national research effort with nearby France and Switzerland. At the current time, there is a tri-national biotechnology teaching program that is centered at the Pasteur in Strasbourg.

This led to a discussion of the role of education in systems biology. Jaeger does not believe in creation of an undergraduate degree in systems biology for now. It is unclear, for example, what an employer could expect from a student trained in such a program. Prof. Reuss thinks there is already too much confusion among the employers. However, he sees some use in the creation of a PhD in the area. Driever believes that systems biology would be an acceptable undergraduate degree but only for the classical disciplines in computer science. He believes they are four to five years away from establishing a degree program in systems biology since the field is immature. Prof. Dr.-Ing. Matthias Reuss said that Stuttgart does have a degree in biotechnology which may be specialized further at the graduate level. Stuttgart has two engineering specializations: one in cybernetics and one in biochemical engineering. In addition they are thinking about constructing a systems biology program. They are aware of the problems, but perhaps a very special small group of students could succeed with a small bit of funding to start things off. They have also started a center for systems biology at Stuttgart with faculty in engineering, mathematics/physics, chemistry/biochemistry, biology/microbiology/immunology, and informatics.

Dr. Eric Karsenti then described his program at the European Molecular Biology Laboratory (EMBL) in Heidelberg. He stressed that systems biology is ultimately about trying to understand how a system maintains itself and reproduces due to the interactions of its underlying parts. This will be accomplished by a well-rendered cycle of experiment, computation and theory. At EMBL they do not have a systems biology department *per se* but they use the whole laboratory as a support for some projects in this "new" paradigm. EMBL has 1,500 people headquartered in Heidelberg but the entire Europe-wide organization is spread to five sites by discipline: structural biology is in Hamburg; sequencing in Hinxton/U.K. at the Sanger Center; basic molecular biology at Heidelberg; more structural biology at the European Synchrotron Radiation Facility (ESRF)/Grenoble (F) and mouse biology at Monterotondo/Italy. These centers are highly interactive and have split the central problem so it can be approached efficiently. For example, Heidelberg is responsible for collecting genetics and biochemical data on the parts of biological systems; Heidelberg and Hinxton are compiling molecular interaction databases; structural genomics is being coordinated from Hamburg and Grenoble. At a systems level, cell morphogenesis and division is a major topic at Heidelberg; synthetic biology (artificial regulatory networks) is being developed at the Society for Conservation Biology (SCB) and Heidelberg; and tools for studying protein and networks are being developed at Cellzome (a proteomics-based praxis drug discovery company) and Heidelberg.

He then went over a number of applications in quantitative genomics. He described a project to use genomics to identify molecules in different aspects of cellular organization and function in HeLa cells. They tagged 4,000 proteins with fluorescent tags, detected and localized them by microscopy, and measured their dynamics by fast microscopy. Reproducibility was improved by robotic liquid handling and automatic image analysis to classify differently behaving clones. He also described a project on elucidation of the mechanisms behind cytoskeletal control of cell shape and division. He demonstrated how models of the molecular motors on mitotic spindle could, in combination with quantitative imaging data, invalidate one of the theories for the directional bias of these motors. This work could be considered one of the successes of systems biology.

Dr. Jens Doutheil from the DECHEMA (the Society for Chemical Engineering and Biotechnology) in Frankfurt spoke next. The DECHEMA is a non-profit scientific and technical society with over 5,000 private and institutional members. He described the administrative structure of the Alliance from the Federal Ministry of Education and Research, which include the international steering committee (see above) and the project committee (Jens Timmer, Matthias Reuss, Ernst-Georg Gilles, Augustinus Bader.) He described in more detail the two Hepatocyte Project Platforms discussed above and how the project was to be coordinated through partnering workshops, annual status workshops, conference organization (5th International Conference on Systems Biology (ICSB), 2004, on October 10–13, 2004 in Heidelberg), contracting and centralization of data management, and an organization for scientific reporting. He went further to describe publicity efforts to engage the international community including flyers and brochures, a web site (www.systembiologie.de), a journalism bureau and the organization of conferences.

Dr. Ursula Klingmüller then spoke on Project B (Hepatocyte regeneration) and related projects. This is a coordinated effort with both the cell biology and modeling platforms. In particular she is working on experimental designs and measurements for parameterization and discrimination of models of hepatocyte behavior with Dr. Jens Timmer at the University of Freiburg. Their goal is to develop data-based mathematical models of pathways involved in regeneration in hepatocytes to predict how to induce proliferation and differentiation.

The effort requires coordination among Freiburg, Heidelberg, Tuebingen, and Wurzburg. Von Weizsäcker at Freiburg is managing murine hepatocyte preparation and standardized cultivation conditions. Klingmüller is focusing on the Jak/Signal transducer and activator of transcription (STAT) pathway and the activation of Smad by transforming growth factor

(TGF)-beta in this cell-type. Merford and Sparna are examining the tumor necrosis factor (TNF) alpha pathways. Mohr is working on the insulin induced ras/mitogen-activated protein kinase (MAPK) pathway. Hecht is working on the Wnt signaling pathway. Bomer is focusing on the Fas-induced caspase cascade for proliferation and apoptosis. Analysis of transcription factors activated by the MAPK pathway is being done by Schutz, Nordheim and Walz. Timmer is handling the modeling side and coordinating with the national platform for modeling. Although the project runs only for six months, the supply with standardized primary hepatocytes is already established.

However, Dr. Klingmüller described an earlier success story on analysis of the Jak/STAT pathway. Here there were two alternative hypotheses about how STAT, after phosphorylation by a Jak-activated receptor, could enter the nucleus and activate gene expression. In one model, STAT5 essentially was trapped in the nucleus after activation and, it was assumed, ultimately degraded. In the other, STAT5 could somehow become unphosphorylated and recycle out of the nucleus and back in again. They generated data on receptor activity by quantitative immunoblotting and measured phosphorylated STAT5. They then used a novel procedure to compare mass-action kinetic models of the alternative hypotheses to the data and were able to demonstrate that without cycling of STAT5 the data could not be explained. They then used the model to predict "unobservable" behaviors of the network.

They then went further to ask at which point in the pathway is it most effective to intervene. The target function was to increase STAT5-p in the nucleus. The conventional view held that increasing phosphorylation would work. Based on their previous model, there was also the possibility of blocking nuclear export. Analysis of the model predicted that blocking nuclear export would be most effective. To test these hypotheses they applied Leptomycin B (which blocks nuclear export) and then measured a STAT5 gene target CIS—and were able to show that nuclear-cytoplasmatic shuttling is an essential systems' property of this pathway (Swameye, 2003).

They are extending both their theory to deal with more complex models and experimental designs and their quantitative immunoblotting techniques to prevent non-specific binding which limits the accuracy of the measurements.

Prof. Dr. Matthias Reuss from the University of Stuttgart was up next. He is working on Project A: A systems biology approach to detoxification and dedifferentiation in hepatocytes. His group works at the Center for Bioprocess Engineering, established in 1993. Currently, he is scaling up for the study of biotransformation of xenobiotics by oxidation, reduction, and hydrolysis by tracking activities of cytochrome p450s, flavones and other metabolites. The effort on this project is organized into a number of different efforts across many universities. Chemoinformatics is being handled at University of Erlangen where, based on the electronic/geometric properties of small molecules, neural networks are being trained to make predictions about which reactions molecules will be processed based on a database of 10,000 drugs and their properties. At the University of Stuttgart there is a molecular modeling and bioinformatics effort in homology modeling and other structural predictions to determine kinetic parameters and to classify different cytochrome p450s. There are also efforts on the kinetic analysis of the metabolic pathways taking into account different environmental factors and polymorphisms. They have started efforts on transcriptome analysis as well—for whole genome and hepatocyte specific arrays and they have developed chips for analysis of single nucleotide polymorphisms (SNPs). Reuss and Mauch have initiated topological analysis and dynamical modeling of the underlying metabolic and regulatory networks. One of their largest models is for the transforming growth factor (TGF)-beta network—which contains 518 reactions. Finally, Michael Dauner is heading metabolome/fluxome measurements and there is a nascent proteomic effort.

Following Reuss, Prof. Dr. Jan Hengstler described his efforts at coordinating the Cell Biology Platform for the Hepatocyte project. The goal of this facility is to supply the consortium with reproducible, quality-controlled, human and mouse hepatocytes. He developed a standardized experimental procedure for quality control. There are three projects to optimize this. Project 1 seeks to develop *in vitro* systems for culturing of hepatocytes. Project 2 is for the development of bioreactors and control of cellular microenvironments. Project 3 is developing a 3D culture system for more realistic hepatocyte development.

The final speaker of the day was Sven Sahle from Ursula Kummer's group at the European Media Laboratroy (EML) Research GmbH, a privately funded research institute focusing on information sciences. This group is coordinating work with Prof. Dr. Gilles at the Max Planck Institute (MPI) by having meetings every six months. They are developing computational platforms such as Sycamore and complex pathway simulator (COPASI) for the building, simulation and analysis of cellular pathways. Their final goal for the program is to build a suite of methods and tools to facilitate the integration of experimental and computational approaches and to support the biological user in the choice of appropriate computational tools to tackle a specific problem.

Overall, the DKFZ visit clarified the integrated structure of the Hepatocyte project and demonstrated a technical ability and mission focus that was laudable. The evident support of the host universities and government in creating integrated research facilities, hiring young faculty in the area, and in developing degree and training programs for the area promises to make this and future projects successful.

REFERENCE

Swameye I., T. G. Müller, J. Timmer, O. Sandra, and U. Klingmüller. 2003. Identification of nucleocytoplasmic cycling as a remote sensor in cellular signaling by data-based dynamic modeling. *PNAS* 100: 1028–33.

Site: **Humboldt University**
Collaborative Research Center for Theoretical Biology
Robustness, Modularity and Evolutionary Design of Living Systems
Invalidenstrasse 43, 10115 Berlin, Germany

Date visited: July 6, 2004

WTEC Attendees: M. Cassman (Report author), A. Arkin, F. Doyle, F. Heineken

Hosts: Reinhart Heinrich, Professor, Theoretical Biophysics,
Tel: +49-30-2093-8698,
Email: Reinhart.heinrich@biologie.hu-berlin.de
Hanspeter Herzel, Professor, Theoretical Biophysics,
Tel: +49-30-2093-9101,
Email: h.herzel@biologie.hu-berlin.de,
Peter Hammerstein, Professor, Organismic Evolution,
Hermann-Georg Holzhutter, Professor, Organic Evolution, Institute of Theoretical Biology,

BACKGROUND

The program visited exists entirely in the Institute of Biology of Humboldt University. The group consists of researchers from the Theoretical Biophysics Department, established in 1992, headed by Reinhart Heinrich from the Theoretical Biology Department. The Institute brings in 140 students per year, about 20 of whom join the Biophysics Department. The Graduate school in Dynamics of Cellular Processes attracts about 15 students per year. There are international graduate collaboration programs with Bioinformatics at Boston University through Temple Smith, the Kyoto Genomics and Bioinformatics Center through M. Kanehisa, and with the BioCentrum in Amsterdam through Hans Westerhoff. These programs include joint workshops, PhD student exchanges and post-doctoral fellowships. They have a well-developed curriculum in bioinformatics, theoretical biology, and biophysics including formal courses on systems biology and mathematical modeling. They also run two-week workshops in the area.

Research Projects

Out hosts were Reinhart Heinrich and Hanspeter Herzel.

Prof. Heinrich laid out the structure of the effort as related above. His own department consists of one full professor (himself), one of the new junior professors (which are young researchers), nine PhD students and two postdocs. Heinrich's major focus has been on dynamic models of metabolism and control of networks but he has since branched out to other sorts of biological pathways. He briefly described work on metabolism, messenger ribonucleic acid (mRNA) translation, metabolic and calcium oscillation, signal transduction in the Wnt pathway, and evolutionary optimization of networks. His approach involves modeling (with close verification by experiments) using methods from nonlinear dynamics and simulation, bifurcation theory, metabolic control analysis, stoichiometric network analysis, stochastic process theory, optimization and graph theory. He sees the flow of information from the structure of a network to simulations of its dynamics to understanding behavior, and from assertions about the needed biological function, available biophysics and efficiency goals to evolved optimal design of those networks. The examples he spoke on included modeling the effect of Wnt stimulation on β-catenin expression in *Xenopus* oocyte extracts that he recently published with Mark Kirschner in the Public Library of Science Journal (PLoS). These models were backed up by careful experiment. He discussed the effect of topology on G-protein signal transduction networks and found that depending on topology there were greater and lesser regions of stability based on both the number of kinases in the network and phosphatase activity. He also discussed forays into the evolution of networks through using the large Kyoto Encyclopedia of Genes and Genomes (KEGG) metabolic maps and asking how such metabolism could be built essentially a step at a time by finding critical routes from a metabolite to other necessary endpoints. Different substrates can make different numbers of primary metabolites through the known reaction network: Adenosine triphosphate (ATP) can make about 1,500 whereas glucose can only make around 50. So a great deal of metabolism could have been extrapolated from a smaller metabolism based only on transformations of ATP. In this work, there were recurrent themes of dynamical modeling, parameter estimation and the effects of network structure along with specific experiments to test the theories. Most students doing this work were physicists and had a long learning curve to understanding the biology. This was in some cases exacerbated by the geographical disparity between the theoretical and experimental groups.

The next speaker was Hanspeter Herzel, a member of the Institute for Theoretical Biology (ITB). The senior personnel at the Institute is composed of Peter Hammerstein (the head) who is interested in the evolution of organismal systems, Andreas Herz who is in computational neuroscience, amd Hanspeter Herzel himself who is most interested in molecular and cellular evolution. There are also junior group members (run entirely on soft money) who include Laurenz Wiskott (neural com-putation), Michal Or-Guil (systems immunology) and Richard Kempter (theoretical neuroscience). The Institute was started on a five-year grant for positions, computers, a small library and some research. Interestingly, five of the 20 professors in biology are theorists. ITB teaches to the entire biology student body and has particular courses in mathem atics and statistics for biologists, as well as theoretical biology A (modeling) and B (evolution). They also teach advanced courses in computational neuroscience, data analysis, nonlinear dynamics, bioinformatics, evolutionary game theory, theoretical immunology and more. There ap-pears to be an in-depth curriculum which is mostly used by those students who are motivated to quantitative and computational approac-hhes.

Professor Herzel's group consists of approximately eight members. There are four immediately proximal collaborators and a number geographically distant, though nearly all the projects were selected so as to have the experimentalist nearby. Though this location does not seem to have any centralized facilities, Prof. Herzel says there are specialists in most of the new technologies to call on nearby.

His current projects include analysis of control in the Ras signaling pathways through the use of dynamic modeling compared with Western blots and microarray data; circadian clock modeling; and a new project on Huntington's disease. In the Ras pathway project they built a large dynamic model (13 equations, 40 parameters) based on previous models and some extensions and predicted that overexpression of Erk would lead to oscillations. Though the experiments did not directly find oscillations they did find complex behavior in Erk. They are now scaling up the microarray time-series and ribonucleic acid interference (RNAi) knockdown work to get better identification of this pathway. In the circadian rhythm modeling he has successfully reproduced wild-type and some mutant behavior of the mammalian oscillator but had to introduce both high cooperativity and a six-hour delay. Exploring a wide class of models at different levels of abstraction, he was able to extract the core features that were necessary to match the data. However, when he compared to Goldbeter's model, which used different mechanisms (MM decay vs. mass action), such strong nonlinearity wasn't necessary. By measuring the decay of certain proteins some of these models might be discriminated. The Huntington project has been frustrating in its complexity. The system itself is large and has many feedback loops but Prof. Herzel is trying to understand the onset of the disease as either arising from these feedbacks or from some sort of nucleation process in the aggregation of the central protein Huntingtin. By using microarrays, and antibodies to aggregation they hoped to find the proximal causes of the aggregation dynamics. However, gene expr-ession studies showed no significant gene expression changes in normal and Huntington's gene expression levels, although specific inhibitors show that there are genes important for the effect, such as Hsp70.

The next speaker was Prof. Hermann-Georg Holzhutter who spoke mostly on the Hepatocyte project. This is a national project which is composed of what appears to be about thirty weakly linked groups studying hepatocytes, including interdisciplinary joint projects on detoxification and dedifferentiation and more methods-oriented projects on cell biology and bioinformatics/modeling. Approximately six groups in Berlin are involved directly in the modeling effort. Their focus is on the development of methods and software tools for the quantitative analysis and complex modeling of cellular networks. This is done in the service of elucidating the design principles of signaling pathways; stud-ying the genetic control of metabolic networks, implementations of vis-ualization of hepatocyte-relevant kinetic models and development of theoretical approaches to include spatial processes in the model of networks. They broke the problem into six modules: calcium mediated cell-cell interaction, cell-cycle control of regenerating hepatocytes, the Wnt mediated β-cantenin pathways, ubiquitin-dependent protein turnover, metabolism and biogenesis of lipoproteins, and intracellular lipid trafficking. They hope to be able to make progress on particular applications such as the prediction of potential oncogenes and tumor suppressors, identification of target enzymes for pharmacological treatment of disorders in the lipid metabolism of the liver, and the prediction of systemic effects upon administration of proteasome inhibitors. The entire project is funded at $30 million for three years. The WTEC team received the impression that little formal infrastructure exists for the generation of standard biological reagents and datasets, computational tools, or even coordination among the participating groups. However, this group was thinking about what the most powerful model-driven experiments might be. Unlike the Max Planck Institute's (MPI) however, there seemed to be little coordinated effort towards all these goals. Rather, there was a commonality of interest in the key questions of how to represent biological processes mathematically, how to simulate and analyze those models and how to compare them to experiments and make predictions of use to a biologist. The structure of this part of their effort became clearer in Heidelberg.

Site: **Max Planck Institute for Molecular Genetics**
Department of Vertebrate Genomics
Ihnestrasse, 63-7314195, Berlin, Germany

Date visited: July 5, 2004

WTEC Attendees: M. Cassman (Report author), F. Katagiri, A. Arkin, F. Heineken

Hosts: Dr. Hans Lehrach, Director,
Department of Vertebrate Genomics,
Tel: (+40) 30 8413 1220,
Email: lehrach@molgen.mpg.de,
Dr. Edda Klipp,
Axel Kowald,
Christoph Wierling,
Dr. Silke Sperling

BACKGROUND

The Max Planck Institute for Molecular Genetics was founded in 1964, focusing on deoxyribonucleic acid (DNA) replication and gene regulation in bacteria and the structure and evolution of ribosomes. In the 1990s there was a major shift in the research orientation of the Institute as new leaders arrived, with a focus on genomics. The Institute has four departments, one of which, the Department of Vertebrate Genomics, is headed by Hans Lehrach. This department has 17 independent groups totaling about 150 people. Activities include sequencing and sequence analysis; functional genomics; and bioinformatics and systems biology. The WTEC team heard presentations from two groups, the system biology group and the cardiovascular genetics group as well as from Dr. Lehrach.

In Dr. Lehrach's introduction, he stressed the need for modeling large-scale biological networks to address the important unsolved problems of disease. This involved bypassing the classical biological paradigm of reducing biology to the study of individual biological entities, and rather producing quantitative models of network behavior that he felt were the necessary basis for understanding complex diseases such as cancer, heart, disease, etc. This approach involves a cyclic process of data generation, modeling, further experimental tests, and refinement of the model to yield predictive outcomes. Three presentations from his bioinformatics/systems biology group described their work in model building and analysis, and one presentation described the work of the cardiovascular genetics group.

RESEARCH PROJECTS

Dr. Edda Klipp is the head of the kinetic modeling group. She described her group's involvement in a variety of collaborative projects including the Berlin Center for Genome Based Bioinformatics; the Yeast Systems Biology Consortium; and the EU project Quantifying Signal Transduction (QUASI), which involves six groups in different EU countries who are quantifying signaling, mostly in yeast. She has two graduate students, and is involved in a graduate program in dynamics and evolution of cellular and macromolecular processes; and a European early stage training network in systems biology. She described the kinetic modeling of the yeast stress response, using the osmotic, pheromone, and starvation stress pathways. The approach is to subdivide the system into modules and generate differential equations to test the models to result in testable prediction. For example, a model and equations were generated for the high-osmolarity glycerol (HOG) phosphorelay pathway in osmoregulation and fitted to known time courses of the key phosphorylation proteins in mutants and other experimental models. Boolean, differential equations, and stochastic modeling were used. A double stress experiment then yielded results that were consistent with model predictions.

The second presentation was by Axel Kowald, a postdoc with Dr. Klipp. He described a model of the aging of mitochondria through free radical damage. A phenomenological model based on differential equations suggested that a model of slow degradation of damaged mitochondria fit the observed data. He is now collaborating with experimentalists in Frankfort to measure some of the predictions of the model.

The next talk was by a graduate student, Christoph Wierling. He has developed an object-oriented tool for modeling and simulation of cellular processes using Python. This can automatically be populated from the Kyoto Encyclopedia of Genes and Genomes (KEGG) database and provides both visualization and simulation. He then took the KEGG map and assumed that the trisomy of the Down's syndrome patient led to an increase in enzyme activity. After running sets of randomly determined parameters through the KEGG model, the predicted stationary states were compared to the normal and the Down's assumptions. Those metabolites which had the largest differences were identified. One was homocysteine, which is known to be differentially displayed in the Down's patient.

Finally, Dr. Silke Sperling is the head of the cardiovascular genetics group who is working on complementary deoxyribonucleic acid (cDNA) analysis applied to normal and malformed human hearts. The purpose of the study is to identify genes associated with problems in cardiac development. She analyzed tissue using high-throughput arrays from various parts of the hearts in normal and dysfunctional hearts (tetrology of Fallot and ventricular septal defects). She was able to see genes differentially expressed and to correlate this with quantitative models of cardiac phenotype. They found evidence for a random chromosomal localization of heart expressed genes with a distribution in two- and three-gene clusters. She is now examining the mechanistic basis for these changes.

Site: **Max Planck Institute for Molecular Plant Physiology**
Am Muehlenberg 1 Science Center Golm
14476 Golm, Potsdam, Germany

Date visited: July 5, 2004

WTEC Attendees: F. Katagiri (Report author), M. Cassman, A. Arkin, F. Heineken

Hosts: Lothar Willmitzer, Director, Dept. 1, Molecular Physiology, Tel: +49-331-567-8200/0,
Email: willmitzer@mpimp-golm.mpg.de,
Mark Stitt, Managing Director, MPI Molecular Plant Physiology, Tel: + 49 (331) 567 8102,
Email: mstitt@mpimp-golm.mpg.de
Wolf-Rüdiger Scheible, Research Group Leader, Molecular Genomics,
Email: scheible@mpimp-golm.mpg.de
Michael Udvardi, Research Group Leader, Molecular Plant Nutrition,
Email: udvardi@mpimp-golm.mpg.de
Victoria Nikiforova, Research Group Leader, System Integration,
Email: nikiforova@mpimp-golm.mpg.de
Joachim Selbig, Professor, Research Group Leader, Bioinformatics,
Email: selbig@mpimp-golm.mpg.de
Dirk Steinhauser

BACKGROUND

An overview presentation was given by Mark Stitt.

Organization

The Institute has 25 independent groups and about 300 people (including administration). It is organized into departments, junior research groups, infrastructure groups, university guest groups, and guest groups. There are three departments: one led by Lothar Willmitzer since 1995, which has seven research groups; one led by Mark Stitt since 2000, which has five research groups; and one that Ralph Bock is starting. The departments have about 160 scientific staff members. Unlike other Max Planck Institutes (MPI), the departments in this Institute are practically merged.

The Institute has a close relationship with the University of Potsdam, which is adjacent to the Institute.

Founding mission of the Institute (1994): To follow an integrated research approach to solve basic questions in plant physiology, combining methods from genetics, molecular biology, chemistry and physics.

CURRENT SITUATION

Informatics efforts were initiated out of necessity to handle a large amount of data generated by wet labs. Thus, the research efforts have evolved into systems biology, which is strongly based on generation of high throughput data in wet labs.

Research Projects

The general experimental scheme is described as "genetic diversity growing in defined environments subjected to broad phenotyping." A variety of experimental design concepts are employed: biased vs. unbiased, single gene vs. multiple genes, alter gene expression vs. alter proteins. Single gene-oriented "rational metabolic engineering" often did not work due to the complexity of metabolic networks. This led to the development of broad phenotyping technology platforms to combine with systematic disruptions of genes involved. More recently, an approach for study of polygenic traits is included in the program.

Genetic Diversity

1. Systematic or targeted over-expression of genes in major metabolic pathways and others
2. High-throughput random gene silencing by antisense ribonucleic acid (RNA)
3. Systematic generation of T-DNA tagged knockout lines
4. Gain access to introgression lines (*Arabidopsis*, tomato, maize, pea)
5. Targeting induced local lesions in genomes (TILLING), which identifies individuals with point mutations in genes of interest from a heavily-mutagenized population.

Phenotyping Technologies

1. Expression profiling

   The Affymetrix array is used for *Arabidopsis*. In-house arrays are used for tomatoes and lotuses. A technology gap was filled by real-time reverse transcription-polymerase chain reaction (RT-PCR) for over 1,400 *Arabidopsis* transcription factor genes.

2. Proteomics

   More than 20 quantitative enzyme assays, which were automated except for the protein extraction step, were developed and used for measurements of protein levels. The assays are based on enzymatic cycling assays (G3P-DAP, $NADP^+$-NADPH, NAD^+-NADH)

3. Metabolic profiling

   Currently, 100 known and 500 unknown compounds can be profiled. *Arabidopsis* is believed to have approximately 20,000 compounds. Multiple platforms, such as gas chromatography time-of-flight (GC/TOF), liquid chromatography (LC), etc., are required for good coverage. Stitt doubts that nuclear magnetic resonance (NMR) would dramatically reduce the number of platforms needed. The relative ratios to isotope standards are used for quantitation. About 20% of total runs are used for controls.

4. Specialized profiling system

   Cell wall profiling is performed by a combination of enzymes that specifically cut sugar chains and matrix-assisted laser desorption/ionization-time of flight (MALDI-TOF).

5. Efforts are being made to run various profiling technologies for different sub-cellular compartments and with a single cell/tissue.

Integrate, Display, and Analyze Data

Interactions between a central bioinformatics group (research and service) and each lab group are facilitated by bioinformatics people in each group. A large bioinformatics tool box for mining and visualization is provided for discovery by biologists (CDB.DB). Interpretation is aided by visualization against the background of known pathways and processes (MapMan).

Single Genes to Multiple Genes

To study polygenic traits, introgression lines are used. Each introgression line carries one chromosomal region of one inbred line in the chromosomal background of another inbred line. Having a collection of introgression lines that collectively cover the entire genome of the former inbred line will help dissect polygenic traits that differ between two inbred lines into monogenic traits. Such introgression lines in tomatoes (150 lines) have been established by Dani Zamir (Hebrew University) and a database including a break-point map and phenotype information for each line is maintained at Cornell University. The MPI group takes part in the international efforts of phenotyping the lines by using its metabolic profiling technology.

Specific projects were discussed in the following presentations:

Victoria Nujufiriva

Discovery of sulfur-starvation related genes based on correlations in expression and metabolite profiles was presented. The threshold for significance was decided by comparing with correlations in shuffled data. A network, including cause-effect directionality, was built based on the significant correlations and known links. Two sub-networks were studied closely.

Joachim Selbig (Bioinformatics)

HARUSPEX, *Arabidopsis* expression database.

MapCave, systematic way to expand annotations.

MetaGeneAnalyse, a suite of analytical methods.

PaVESy, pathway visualization and editing (KEGG is generic, need to specialize it for plants).

PDM (plant diagnostic module), supervised learning for diagnostics. Decision tree machine was used because it is easy to grasp the decision-making process used.

For community use, the software was made easy to use, and in this way, more people use the software. Consequently, more feedback can be obtained from users. The small size of the research community helps in this respect.

Dirk Steinhauser

Comprehensive systems biology database (CSB.DB)

It holds publicly available expression profile data from different organisms. It allows co-response query and returns a functional category summary. This helps identify candidate genes, which can be further analyzed using CSB.DB, including use of MapMan, which is a functional category-classified expression viewer. Two questions were raised: 1) How should particular software be compared with other similar programs, and 2) what is the best strategy for a research community to deal with competing developments?

Wolf-R Scheible

Forward genetics had not been very successful with nitrogen-regulation studies due to functionally duplicated genes (recent duplication of the genome is common in plants). Therefore, a reverse genetic approach was taken. Their real-time reverse transcription-polymerase chain reaction (RT-PCR) platform for *Arabidopsis* transcription factor genes quantitates >1,400 genes (there are ~2,000 transcription factor genes in total) with >90% reliability with one transcript/1,000 cells sensitivity, which is much better performance than an Affymetrix array can achieve. The development of the platform required ~$300,000 initial investment. Forty nitrogen-regulated genes were identified and analyzed by inducible over-expression.

Mark Stitt presented further details of enzyme activity profiling and MapMan. Enzyme activity profiling was used to measure protein level changes during the day, and the protein level changes were compared with the corresponding mRNA changes. Generally, no clear correlation between mRNA and protein level changes were observed. The degree of correlations between mRNA and protein levels varies in different settings. Thus, measuring mRNA and protein levels in each setting is important.

Budget

The Institute receives approximately €8 million for everything except depreciation. Approximately €3.5 million of this and additional approximately €3.5 million from competitive grants are used as the research budget. Generally, the overhead rate is zero or up to 20%.

Education

Most bioinformaticians begin as computer scientists. To develop computer scientists into bioinformaticians, the motivation to take up biological studies is an important factor. Germany should have initiated efforts in bioinformatics education earlier. MPIs cannot influence college programs, which is an unfortunate situation since MPIs are often leading research institutes in the country. Another challenge in changing education programs at the college level is that universities cannot make decisions about curricula by themselves—new curricula need to be approved by higher governmental organizations.

Site: **Oxford Brookes University**
School of Biological and Molecular Sciences
Oxford OX3 OBP, U.K.

Date visited: July 6, 2004

WTEC Attendees: D. Lauffenburger (Report author), C. Stokes

Hosts: David Fell, Tel: (+44) 1865 483247, Email: dfell@brookes.ac.uk

BACKGROUND

Central theme:

Models of Cellular Processes

1. Metabolism—plants, bacteria; for biotechnology
2. Cell cycle—human cells; for therapeutic

Research Projects

- Threonine biosynthesis in *E. coli*
- Potato tuber metabolism
- Collaboration with Advanced Cell Technologies in Cambridge
- Photosynthesis
- Collaboration with Christine Raines at Essex University
- Actinomycete antibiotic production
    1. Collaboration with John Ward at University College London and Peter Leadlay at Cambridge University
        a) Funded by Functional Genomics Initiative of Biotechnology and Biological Sciences Research Council (BBSRC) (There is now coming an analogous new BBSRC program, the Systems Biology Initiative, perhaps similar to National Institute of General Medical Sciences (NIGMS) Centers of Excellence in Complex Biomedical Systems.)
            - Overall program is funded at about £6–10 million/four years, though, according to Dr. Fell, BBSRC claims the amount may be closer to £25 million, but the real figure is not known
            - Fell/Ward/Leadlay project has about £900,000 in funds for four years

Startup Company

- Physiomics
    1. Software permits modeling of multiple individual cells simultaneously, including appearance of new cells as they arise from cell division
    2. Individual-cell model focuses on cell cycle control (G1/S transition)
        a) About 100 components (cyclins, cyclin dependent kinases (CDKs), etc.)
        b) Ordinary differential equation (ODE) framework
        c) More detailed than Tyson approach, which emphasizes simplifications for gaining intuitive insight; seeks to be able to predict and test effects of specific molecular target inhibitors
        d) Collaboration with Cyclacel, a Dundee-based company that has cell cycle-related potential therapeutic compounds
        e) Data are currently literature-gleaned, but intend to start commissioning dedicated experiments for obtaining protein levels (and perhaps kinetic rate constants)

Perspective

- New U.K. government funding agency push toward bringing together previously disparate communities—high-throughput data-driven genomics/proteomics community, and model-centric "physiological function" community
- Ten-year vision to establish 'predictive biology,' or 'integrative biology'
- Metabolic engineering approaches work for small-scale (~10 component) systems with dynamics, or larger-scale (~100–1000 component) systems with only stoichiometric structure
- Technical challenges
- Cell cycle is more difficult than metabolism for kinetic modeling, due to much less usable experimental data
- Cell function modeling generally neglects downstream biophysical processes involved in executing function regulated by upstream biochemical pathways
- Packaging larger individual-cell models into multi-cell simulations
- Structural modeling—deconstruction of large networks into smaller substructures

1. Elementary Mode Analysis
    a) Predicts pathways feasible from gene expression information
    b) Predicts pathways capable of greatest metabolic yields (mass, energy)
    c) Major success story: prediction that there should be six categories of arid-environment plants in terms of crassulacean acid metabolism
        - Previous dogma was that there were four
            I. One type had been previously known as not fitting into these, but ignored (*Aloe vera*); thus, fifth
            II. Another type has been recently discovered (*Mesembryanthemum crystallimum*), as sixth

Education

- No new educational programs or curricula planned
- Students can already do double major in computer science and biology

Site: **Oxford University, Centre for Mathematical Biology/Mathematical Institute**
24-29 St Giles', Oxford OX1 3LB, U.K.

Date visited: July 5, 2004

WTEC Attendees: C. Stokes (Report author), D. Lauffenburger, H. Ali

Hosts:
Philip Maini, Director, Centre for Mathematical Biology, Tel.: 44-1865-280617,
Fax: 44-1865-270515,
Email: maini@maths.ox.ac.uk
Jon Chapman, Professor, Oxford Centre for Industrial and Applied Mathematics (OCIAM), Mathematical Institute
Chris Scofield (experimental chemist),
Jotun Hein (bioinformatics, statistics and genetics)

BACKGROUND

Institutional Context

Centre for Mathematical Biology (CMB) was founded in 1983 to foster interdisciplinary research in mathematical biology. Historically they've had about six grad students and one to two postdocs per year although currently there are about 20 grad students and postdocs. Most students are co-supervised by a life scientist. Many areas of research are addressed, including wound healing, pattern formation, and cancer.

Oxford has applied for funding (£6 million over three years) from the Biotechnology and Biological Sciences Research Council (BBSRC) for a new center in systems biology. If funded, Oxford is committed to creating space for the group in a new building and hiring five new lecturers.

Some new courses have recently been introduced addressing systems biology. They also have a five-year doctoral training center (DTC) for the life science interface. Students receive a four-year fellowship; in the first year they do courses in areas they don't know (bio for physical scientists/math/engineers, the reverse for life scientists). Fifty studentships will be supported in total.

They are finding significant interest now for the use of modeling in biological research, with many biologists seeking them out. Uniquely, they hired a "bio liaison officer," a post-doc, who spends 50% of his time networking with life scientists and industry. He's been so successful matching people up from both the university and industry that they've slowed his work because they can't engage in all the possible projects. They have two projects with companies– one on a gel for wound healing and one for a bioreactor.

RESEARCH PROJECTS

European Network Team Studying Cancer

Dr. Philip Maini described a collaboration of 10 teams across Europe focused on multi-scale modeling of nutrient delivery to tissue, regulated by vasculature, particularly in cancer. The project is for three years funded at €3 million from the "Sixth Framework," with one post-doc at each site.

They are addressing how the vascular network responds to physical, chemical and metabolic cues in the tissue and blood to bring nutrients to the tissue, with O_2 the first nutrient of interest. They use models to see how the vasculature will remodel over time, and the consequences of that for nutrient delivery. A mix of techniques is used to describe the various processes; ordinary differential equations (ODEs) for blood flow, partial differential equations (PDEs) for diffusion in the tissue. Cells are individual automota distributed in the tissue, with the use of O_2 by each cell determined by its current activities (e.g., replication) described with ODEs (for intracellular processes). They describe their model approach as "middle out"—start at cell cue-response level, then go deeper to intracellular mechanisms that drive the responses, and also build multi-cell and tissue models to connect cell function and intracellular signaling to physiological outcomes.

One prediction is that the heterogeneous O_2 concentration arising in the tissue because of cellular and spatial differences can greatly influence cancer cell growth. In other work they are analyzing cell cycle models to understand why the cells respond as they do under hypoxic conditions (cancer cells may become quiescent instead of apoptosing). An interesting result was just published in J. Theor. Biol. They used the model to show how regulation of p27 might be different in cancer cells and how it can account for why these become quiescent rather than apoptose like normal cells would. The result hasn't yet been confirmed experimentally.

Another focus is to analyze drug delivery to tumors; doxyrubicin has been simulated. One difficulty is a lack of parameter values, so getting quantitatively specific answers to help plan clinical protocols is not yet possible. They hope their Israeli colleagues will make some of the needed measurements.

E-Science

This project encompasses universities from a number of different countries. The U.K. representatives are Oxford, Nottingham, Birmingham, and

Sheffield. The goal is to use electronic science to attack medical problems, with a focus on heart and cancer modeling. At Oxford, Denis Noble and others (unspecified) are involved.

For the heart, the focus is on using efficient numerical techniques to couple Noble's models of heart function and cell details with mechanical models from Auckland.

For the cancer project they have a similar goal, but models are not so well developed yet. They're creating models of tumors; specifically, a model of a cancer cell including internal molecular processes (cell cycle models) and processes at the tissue level, such as delivery of drugs via vasculature. The focus is colon cancer. They're collaborating with the Institute of Molecular Medicine on hypoxia inducible factor (HIF) signaling that results from hypoxia. HIF is involved in many cell processes including expression of 300–400 genes and numerous competing/opposite effects. They will use models to try to elucidate what's happening when HIF is upregulated and what would happen if HIF or a downstream molecule is targeted for treatment modalities. They've selected four main downstream pathways from HIF as likely important to many of its effects. The experimental part of the project is to measure what happens when elements of these pathways are modulated using standard molecular biological and cell biological experiments.

Another part of this is collaboration with computer scientists to create methods for integrating models. While they're still in the requirements stage, they want to find ways to (a) make it easy to have multiple users of a model, (b) easily expand a model and (c) integrate multiple models to allow for swapping in/out different models, have different types of models talk to each other, etc.

Histidine Sensitivity Research of Judy Armitage

The goal is to create a comprehensive molecular description of all pathways involved in histidine sensing in bacteria (using *E. coli* and *R. sphaeroides*). This project takes a bottom-up approach: collect "all" the data, create a model and see if it can reproduce known system behaviors.

Hein Area of Research: Bioinformatics, Specifically Population Variation

Dr. Hein uses coalescent theory to understand sequence data of populations. He also works on molecular evolution and comparative genomics. One goal is to get information on structure from gene evolution, just as people have used it to get information on sequence.

TYPES OF MODELS

This group uses a variety of model types depending on the problem of interest. They use both detailed intracellular network models (relatively comprehensive) as well as "reduced" or "caricature" models (small abstractions of the more detailed models) as needed to answer the questions of interest, at least by what's possible with the available data. They have an interest in developing comprehensive molecular models of certain systems, but find the data not yet available. They have a preference to simplify detailed models based on asymptotic analysis but often don't have the quantitative data to do so and must rely on biologists' knowledge of relative magnitudes/rates to decide what's in/out. Part of the e-Science project is to actually build multiple sorts of models of the same thing to see where particular types of abstractions are appropriate and where they are not. They support doing modeling now regardless of whether all the data is available for use in understanding the biology and to focus further experiments. One obstacle that was noted is that one doesn't necessarily know in advance what level of detail or even what type of model is most appropriate to address a problem of interest. Another is a lack of experimental techniques to measure all the things that need measuring.

ADDITIONAL INTERESTING MODELING RESULTS

In addition to results noted with projects above, several other notable modeling studies were discussed. They described how Noble's cardiac model had illustrated counterintuitive effects of blocking certain ion channels. One would expect blocking it would be good for certain heart conditions, but the model shows the opposite because of the complex network involved.

In other work on wound healing, experiments show that manipulating levels of transforming growth factor beta (TGFb) decreases dermal scarring, but since TGFb has many effects it's hard to know how it's doing so. The reason scars get worse is that more collagen is laid in parallel instead of across. A wound healing model was used to examine the various TGFb activities and predicted that the most crucial effect of TGFb is that it causes the cells to migrate more randomly and thereby lay collagen down in a more random manner. This was published in Wound Regulation and Repair a couple of years ago. No one has yet confirmed this prediction experimentally.

Site: **Oxford University, Dept. of Physiology Parks Road, Oxford, Ox1 3PT U.K.**

Date visited: July 6, 2004

WTEC Attendees: C. Stokes (Report author), D. Lauffenburger, H. Ali.

Hosts: Denis Noble, Tel: 011-44 (0)-1865-272-533 (554 Fax), Email: denis.noble@physiol.ox.ac.uk
Peter Kohl, Tel:011-44 (0)-1865-272-114 (554 Fax), Email: peter.kohl@physiol.ox.ac.uk
Ming Lei, Tel: 011-44 (0)-1865-272-560, (554 Fax), Email: ming.lei@physiol.ox.ac.uk
Penny Noble, Tel.: 011-44 (0)-1865-272-533 (554 Fax)

BACKGROUND

Cardiac Program

Dr. Noble has been studying cardiac biology through experimental and modeling work for over 40 years. His program aims to connect levels of biology, from molecule to cell to tissue to organ. Over time he has collaborated with numerous researchers, notably Raimond L Winslow, Peter Hunter, and Andrew McCulloch. Early work with Winslow involved creating a multicellular cardiac tissue model with each cell calculated separately and with cell-to-cell connections included. This is very computationally intensive and impractical for a large piece of tissue or the whole heart. Current models use finite element calculations, where each grid point represents a piece of tissue that has homogeneous cell properties attributed to it. Hunter's group has provided a framework for these simulations (continuum mechanics, image analysis, signal processing and system identification (CMISS)). Important anatomical information was obtained from imaging work in McCulloch's lab. With this model they can simulate waves of excitation in the heart, specifically the dog heart. They find that the cell-level models and fiber structures impact results. They use data from their own lab and from many other places to develop and test the model.

What's made the extensive multi-level model of the heart possible is that relevant experimental work (electrophysiology in particular) has been ongoing for 40 years, providing a vast body of data and knowledge. In addition, the major regulators of the cell and tissue function (ion channels generally) are quite accessible to measurement, and there are relatively few cell properties that contribute to whole organ function and which don't depend on vast intracellular signaling networks.

The model doesn't deal with progression of disease in the heart dynamically. Rather, the user can impose differences in parameters (e.g., different wall thickness) to look at snapshots in time if they want to explore some disease in which progressive changes might occur. Dr. Noble referred us to Natalia Trayanova at Tulane and her work on defribillation for some chronic work.

A key requirement for successful modeling is the continuous iteration between theoretical and practical experimentation. Experimental data feeds the modeling effort, and experimental research is required to validate model predictions (Kohl et al., 2000). This is the bedrock of Noble's modeling philosophy.

RESEARCH PROJECTS

Peter Kohl described his research on mechanical effects on the heart. In one project, he used modeling and experimental work to understand how a particular medical procedure works, specifically, thumping the chest strongly for restarting a stopped heart or slowing down one that is beating too fast. He showed a simulation of 250x250 neurons as a 2D slice through the ventricular wall. He found that mechanical stimulation (through stretch receptors) in a certain area at a certain time of heartbeat can result in arrhythmia, while other times/areas of stimulation wouldn't (Sideman, 2004). This reproduces what is seen experimentally. However, they found that the mechanisms by which the arrhythmias arose were quite different than what was expected, so they've gained some insight into underlying biology. They've not yet verified experimentally the latter findings. In the lab tour, Dr. Kohl illustrated how they are measuring the mechanical effects on the heart beat. In a related clinical study, they measured how physicians in U.S. and Britain did the chest thump (speed, force) and found specific differences that were related to greater success in the U.S. in saving patients. He's now following up to influence training in the procedure to improve its effectiveness.

Ming Lei also gave us a tour of his lab and described research on several posters. He's doing gene expression of the mouse heart and looking for ion channels that are differentially expressed in different parts of the heart, especially the atrio-ventricular (AV) node. After gene expression, they then work to understand how expression differences could relate to the function of the AV node compared to other areas. One issue discussed was that messenger ribonucleic acid (mRNA) levels don't relate well to protein expression, and so it's difficult to know how directly relevant the mRNA measurements are to functional studies.

Dr. Noble noted that while they've been quite successful in linking several levels of biology from ion channels to organs within models and learning from those, they have not yet progressed much in his lab to linking to more detailed molecular levels, notably, the electrophysiology to metabolic and genetic pathways. On this he noted that the WTEC panel should investigate the group at Kyoto University led by Prof. Akinori Noma (Physiology), who, with ¥7 billion in investment, heads a consortium of five universities on biological simulation.

They have used and/or discussed using their models with pharmaceutical companies for drug development, notably Novartis and GSK. An abstract by Helmlinger at http://www.bc2.ch/2004/ abstracts_list.html #Helmlinger (Basel computational conference) describes a project with Novartis. For certain chemical compounds, Novartis knew some functions and their whole cell effect but did not have a full functional characterization, so the group used a model to infer the other effects the compounds were likely having rather than going and doing all the experiments it would have taken to narrow it down.

The project direction for the next five years is to work on arrhythmia at the whole organ level. They have significant work now at the cellular level. In addition, the group will continue to further develop and make available enabling software/technology (COR, GRID, CellML, AnatML, FieldML, etc.)

Dr. Noble noted that a large cardiac research effort exists at Oxford beyond his laboratory. Others he noted include Richard Vaughan-Jones, who studies proton transport, which is important in ischemia, both experimentally and through modeling; and Kieran Clarke, who has a large nuclear magnetic resonance (NMR) team working on ischemia. He noted that Peter Hunter is taking a visiting professorship at Oxford and will be in residence for two to four months per year.

This group has a well-established program of both experimental and modeling work. They noted that doing their own experimental work to test their models was critical to their success using models, since they gained significant insight from the interplay between experimental and modeling work.

Lung Research

Dr. Noble only briefly described their work on the lung. For this program, Hunter in New Zealand is again contributing anatomical work, and there is experimental and modeling work also being done with collaborators in Bordeaux.

Systems Biology

The group discussed a number of ideas about systems biology. Dr. Noble posited that systems biology is "the" post-genome science, noting, however, that proponents need to be careful about how the field is sold. The necessity and promise must be described, but the difficulty must not be overlooked. Dr. Noble's view is that a fully bottom-up approach—whereby "all" the data is obtained prior to attempting to model a system—is not appropriate (see summary in Nature's Encyclopedia of Life Sciences). His rationale includes the argument that the function of proteins is not fully specified in the DNA but rather emerges at multiple levels of organization; the combinatorics of including every protein and interaction is simply too large to do computing on even if that data is obtained; furthermore, there are feed-forwards and -backs across biological levels, so knowing only information from the genome will not be sufficient anyway (see also Noble, 2002). This view of "not needing all the data first" isn't necessarily the generally held view by funding agencies or researchers in U.K., he reported, although funding agencies are funding more and more modeling/quantitative work now.

Dr. Noble's approach is to start somewhere in the middle of the biological hierarchy, where a lot of data is available, and work both ways up to physiology and down to genes. He noted that for any system, we won't know *a priori* exactly what level to start at. A key aim is to find what level (gene network, cell, tissue…) is required to assign a function; e.g., some functions can be described by a network of a few genes, some only at the cell (e.g., an action potential), some only by groups of cells (e.g., traveling waves of excitation in a tissue).

SOFTWARE RESOURCES

Dr. Noble's lab is investing significant effort into creating software that can improve model integration and communication by modelers. He notes that a major problem is that many models are not published accurately, and even if accurate, it requires significant effort for another researcher to translate the equations into software to use such a model or to integrate it with another model. He is involved in several efforts to address such problems, including the development of CellML and an associated model repository (see http://www.cellml.org), as well as the development of canonical correlation analysis (COR), a model development and simulation package (Garny, 2003). The first version should be available to others in the near future.

Funding/Resources

Dr. Noble noted that funding is increasing in systems biology in the U.K., with more interest in quantitative and modeling work starting with the Engineering and Physical Sciences Research Council (EPSRC) and the Biotechnology and Biological Sciences Research Council (BBSRC). The U.K. government is involved in the development of grid computing for the U.K. via EPSRC and Department of Trade and Industry (DTI) initiatives. He noted that the grid computing initiative has pulled the computer science (CS) people in and should result in the availability of more powerful computing for projects such as his.

Education

Oxford has significant interdisciplinary programs related to biology. There is very good integration of mathematics, computer science, engineering, physiology, and clinical sciences. Some of Noble's models are used in the medical school to teach medical and physiology students. They've found the students accomplish more with the models than when they've tried to get everyone to do a patch clamp in the lab.

REFERENCES

Garny A., P. Kohl and D. Noble. 2003. Cellular Open Resource (COR): a public CellML based environment for modelling biological function. *International Journal of Bifurcation and Chaos* 13: 3579–3590.

Kohl P., D. Noble, R. L. Winslow and P. Hunter. 2000. Computational modelling of biological systems: tools and visions. *Philosophical Transactions of the Royal Society A* 358: 579–610.

Noble, D. 2002. The rise of computational biology. *Nature Rev Mol Cell Biol* 3: 460–462.

Sideman S., 2004. Cardiac Engineering—Deciphering the Cardiome. *Ann NY Acad Sci* 1015: 133–143.

Site: **Sheffield University**
Computational Biology Research Group
Functional Genomics
Division of Genomic Medicine
Computer Science Dept, Regent Court 211
Portobello St., Sheffield S1 4DP,
University of Sheffield
Beech Hill Rd., Sheffield S10 2 RX, U.K.

Date visited: July 8, 2004

WTEC Attendees: C. Stokes (Report author), D. Lauffenburger, F. Katagiri

Hosts: Chris Cannings, Professor, Division of Genomic Medicine, Tel: (+44) 114-271-2398, Email: s.dower@sheffield.ac.uk,
Richard Clayton, Senior Lecturer, Bioinformatics, Tel: (+44) 114 22 21845, Email: R.Clayton@dcs.shef.ac.uk
Will Zimmerman, Senior Lecturer, Chemical and Process Engineering, Tel: ++ 44 (0)114 222 7520, Email: w.zimmerman@shef.ac.uk
Eva Qwarnstrom, Head, Academic Unit of Cell Biology, Professor, Cell Biology, Tel: 0114 271 3181, Email: e.qwarnstrom@sheffield.ac.uk
Nick Monk, Lecturer, Computer Science, Tel: (+44) 114 222 1832, Email: n.monk@dcs.shef.ac.uk
Steve Dower, Professor, Email: s.dower@shef.ac.uk
Francis Ratnieks, Professor, Department of Animal and Plant Sciences, Tel: 0114 2220070, Email: F.Ratnieks@Sheffield.ac.uk
Rod Smallwood, Professor, Medical Physics and Clinical Engineering, Email: r.smallwood@sheffield.ac.uk
Mike Holcombe, Professor, Computer Science,
Phillip Wright

BACKGROUND

The WTEC attendees spent a day at the university and heard numerous short talks on research programs and engaged in open discussion about systems biology and research programs at the university. Slides from all the talks but Monk's were provided to us. Individual affiliations were given on a handout.

Institutional Context

This group is not a formal systems biology group but happens to have a number of collaborations among them along systems biology lines. There's no specific organization at the university related to systems biology. A formal but unfunded "Center for Bioinformatics and Computational Biology" group exists. Engineering has traditionally been strong at the university, and recently biological departments have caught up, and collaborations like those discussed during the visit are becoming common. In the U.K. there's no real building of large groups in systems biology as standalone departments or centers.

They are submitting a proposal to the national call for systems biology research center grants on the topic of host-pathogen interactions. More than 20 faculty members are investigators on the proposal, which will request £7 million over five years. The funding agency commitment is for three center grants; there may be a second round in the future. University research funding as a whole is about £100 million per year; about 10% of which is for interdisciplinary projects.

There are no formal systems biology education programs, although some undergraduate courses in computer sciences are related to systems biology, and the group has discussed a proposal for an MS in computational biology. It was noted that there is a shortage of graduate students who enter with the "right background" to do truly quantitative work in biology.

There's not much interaction with industry in the systems biology area, except Phillip Wright's work related to the proteomics technology development. The University of Sheffield has significant industry interaction in engineering, medicine, and biology. Probably a third of university funding is from industry (much in engineering and software). The university owns a successful technology transfer company (~15 people) dedicated to evaluating opportunities for intellectual property. The universities are expected to get about a third of their research funding from sources other than the national government and private charitable groups; this means industry, tech transfer, and regional development boards.

Individual Research Areas

Slides from each of the talks were provided, so only brief descriptions are provided here.

Smallwood: Modeling tissue development, using agent-based models, with a focus on how individual cells organize into tissues. Common features include self-assembly, forces between cells, cell motility, mechanotransduction. Have experiments and models. This work is associated with the international Physiome Project.

Ratnieks: System organization in insect societies. The most successful approach is using experiment and theory in tandem. Put forth idea of social cancers—individual bees don't kill the eggs of the worker bees, so they hatch and become female workers, but they don't do work, so the colony breaks down. Good success in learning how ants "find their way home" useing pheromone trails (study of direction and geometry).

Cannings: Interests are math models of how complexity and emergent behavior occur and the generic properties of biological systems. Described project on analysis of protein-protein interaction networks. Experimental data used are yeast two-hybrid results by Uetz et al. and Ito et al. He showed that a domain model (which is based on an assumption of a network generation mechanism) by Thomas et al. explains the data better than the power law (which only describes the organization of a network). Discussed the point that just because a model fits data well does not necessarily reveal a correct underlying mechanism. Eventually would work on how network topology arises and how genetic variation (mutations) could influence that.

Qwarnstrom: Signal transduction for inflammatory mediator signaling (esp. NFκB pathway). Very detailed single cell work—examine how individual cells vary compared to what is measured in a population. How to analyze? Asked what she'd use models for if she had them, she suggested it might help her translate her altered experimental systems to the physiological situation (although she uses low transfection levels to adversely perturb the system as little as possible).

Holcombe: Validating agent-based models of systems biology. Collaborates with Ratnieks on insect societies and with Qwarnstrom on modeling NFκB signal transduction. Used agent based models to track the various signal molecules; found it a good way to deal with the spatial heterogeneity within a cell.

Dower: Immune and inflammatory responses. Interested in toll receptor family. Thinks of cells as computational devices: many cues ➔ intracellular programs interpret ➔ various responses. Building up description of network. Have set up high-throughput robotics platform to work on it. Ignores the dynamics of the systems so far. His conclusions (although still speculative) are that dedicated pathways to specific behaviors are unlikely; rather, combinations of signals specify behaviors. Also thinks the signaling network may essentially be binary.

Wright: Focus on proteomics. Developing new methods as well as study some bacteria including extremophils (sulfolobus). Focus is on how to determine relations between genotype and phenotype, esp. dynamic system focus.

Zimmerman: Used the frequency and amplitude response approach to microfluidics study and kinetic models of metabolic networks. For the latter he particularly talked about yeast glycolysis, where a model was built by Pritchard and Kell. Collaborating with Wright on models of sulfolobus.

Monk: Cell signaling and pattern formation. Studying how to go from a topological network like biochemical pathways and learn how the network actually functions. "How do we represent the arrows" in the network, especially quantitatively. Lateral inhibition of bristle spacing as related to delta-notch is one project. Need for delays in activity of protein interactions was a major result which contradicts another's (Odell) work. Conclusion: need to recognize and study regulation occurring through protein interactions, not just at translational steps. Another point—need to validate modeling approaches against various sorts of problems to see which are really appropriate for which problems.

Clayton: Ventricular fibrillation work. Integrative model of the heart, including putting the heart into a torso to simulate an electrocardiogram (ECG). Nice integrative work at more physiological levels. He noted that integrative approaches including modeling in cardiology (in the field, not only his own work) have led to the ability to define instabilities that can lead to fibrillation and that's being used now to help pharmaceutical development.

Funding

The group confirmed comments from other universities that a major drive for systems biology and interfacing of life sciences with engineering and physical sciences comes from government funding agencies, even more than from the research community. Engineering and Physical Sciences Research Council (EPSRC) was an earlier driver than Biotechnology and Biological Sciences Research Council (BBSRC), and Medical Research Council (MRC) has been slowest.

Nonetheless research funding was generally noted as a difficulty. One particular situation noted is that in England it's quite easy to get a little funding for a short time for a new innovative project without preliminary data (one postdoc funded for three years). But once one has the initial proof of concept and needs to expand (three to four postocs, several million pounds) it's nearly impossible to get that level of funding, so lots of things die out or get taken to fruition by others outside the U.K.

Risks Noted

The actual benefits of doing systems biology are still unclear, so it's perceived as risky. In addition, the U.K. "grading" of faculty and departments every five to six years discourages people from starting in new fields because it could pose a risk to one's ratings if they weren't successful in publishing in top journals. That rating is only a small amount of a department's funding now, however, and may go away altogether after the next round in a year.

Site: **Systems*X*—A New Systems Biology Institute in Switzerland**
Physical Electronics Laboratory
ETH Zürich, Hoenggerberg,
HPT-H6 CH-8093, Zürich

Date Visited: June 29–30, 2004

WTEC Attendee: M. Cassman (Report author)

Host: Henry Baltes, Professor, ETH Zürich
Tel: +41 1 633 20 90, Fax: +41 1 633 10 54
Email: baltes@phys.ethz.ch
http://www.iqe.ethz.ch/~baltes

BACKGROUND

The ETH Zürich, the University of Zürich, and the University of Basel have generated a collaborative project entitled "Systems*X*," which is intended to serve as a focus for systems biology in Switzerland. The structure will accommodate collaborative efforts across disciplines and locations. To ensure integration between the several sites involved the project management will be at the highest levels, comprising the president of the ETH Zürich, the rector of the University of Basel, the rector of the University of Zürich, the vice-president of research at ETH Zürich, the Zürich vice-rector of research at the University of Basel, the pro-rector for research at the University of Zürich, research representatives from Novartis and Roche, plus the spokesperson for Systems*X*. Components of Systems*X* will include the Functional Genomics Centre Zürich, the Glycomics Initiative at ETH Zürich and the University of Zürich, the Oncology Cell Transfer Project at the University of Zürich, the Basel Bioinformatics Initiative, and the Life Sciences Training Facility at the University of Basel.

The organizing principle for the science is that of interdisciplinary clusters, which may be shared among institutions and locations. These clusters will be organized "bottom-up" by research faculty and based on shared research goals. They will be funded by Systems*X*. Clusters will be determined in several ways. Proposals can be developed by the participating faculty. In Zürich, a major proposal in the area of metabolism has been submitted for consideration. Additionally, a research concept can be determined by institutional groups that define strategically important areas. Participating faculty can be identified based on their interest in developing programs relevant to the Institute, or from a frame of predetermined topics for which faculty are to be selected.

Two major initiatives are linked to Systems*X*. They are a new ETH Zürich Center of Biosystems Science and Engineering (C-BSSE) in Basel and a Cluster of Biosystems Science (CLU-BSS) in Zürich. The C-BSSE will be an autonomous Centre of Excellence with its own positions, space, and money. In general, faculty will be affiliated with a department at ETH Zürich, and they may also hold joint appointments at the Universities of Zürich or Basel. However, they will be housed in the C-BSSE building in Basel. It is expected that research groups in systems biology in Basel, (such as the university hospital, Roche, and Novartis), will be networked with faculty at C-BSSE. Funds have been requested by the University of Basel from the Swiss Education Conference that will help support innovative research and teaching in systems biology. Initially, four new professorships at the Basel Biozentrum will be linked to this effort.

The second initiative, the Cluster of Biosystems Science (CLU-BSS) in Zürich will focus on the areas of proteomics and transcriptomics related to tumor research. The leader of this new effort will be Dr. Rudi Aebersold, a leader in the area of mass spectrometry as applied to proteomics. Eight new appointments in areas linked to systems biology will be made by 2007 at ETH Zürich and 10 at the University of Zürich. The faculty of the Zürich grouping will be housed in existing space at ETH Zürich and the University of Zürich.

Support for Systems*X* is being provided both by federal and cantonal funds. C-BSSE will require 33 million Swiss franks (CHF) for construction and operation. Of this CHF 20 million was given by the canton of Basel, 5 million by the Swiss Universities Conference (CUS), and the remaining 8 million will be raised by ETH Zürich. Almost all of this will be for operational support of administration, new faculty, students, post docs, in addition to seed money. Another 3 million is being provided by the CUS for the CLU-BSS in Zürich and 2 million for the corresponding efforts at the University of Basel.

Site: **University College London**
Wolfson House 4 Stephenson Way
London NW1 2HE, U.K.

Date Visited: July 5, 2004

WTEC Attendees: D. Lauffenburger (Report author), C. Stokes, H. Ali.

Hosts:
Anne Warner, Director of CoMPLEX, Department of Physiology, Tel: (+44) 20-7679-7279, Email: a.warner@ucl.ac.uk
Jonathan Ashmore, Department of Physiology, Theoretical Physics Training
Anthony Finkelstein, Computer Science Department, Tel: (+44) 20-7679-7293, Fax: (+44) 20-7387-1397, Email: A. Finkelstein@cs.ucl.ac.uk
Robert Seymour, Department of Mathematics

BACKGROUND

CoMPLEX (Centre for Mathematics in the Life Sciences and Experimental Biology)

- University College London (UCL) umbrella organization for systems biology
- "Virtual department"
- About 100 faculty members university-wide, by self-association
- Also related to Bioinformatics Center (how?) and physics interest in molecular-level measurement (how?)
- Seeking multiple large multi-investigator grants
- Runs Medical Research Council (MRC) doctoral training program for interface between life science and physical science
  - Mainly math/physics background
  - Curriculum includes six to eight "case presentations" during the first year of study, with investigators engaged in multi-disciplinary collaborative research demonstrating to students how the projects operate
  - Experimental work is required
  - Students have two thesis advisors, one from life science and one from physical science

- Also physically houses postdocs working at life science / math-physical science interface

Major Example Project

- Supported under Department of Trade and Industry (DTI) BEACON program (direct funding from DTI is unusual; most funding goes through research councils [e.g., MRC, Biotechnology and Biological Sciences Research Council (BBSRC), etc.])
- "Virtual Integration Across Biological Scales"
- One of seven current projects around U.K., total of approximately £8 million over five years
- Goal is to build *in silico* model of liver (mainly metabolism)
  - Chosen because of pharmaceutical interest and large basis of experimental data in literature
  - Focus on glucogenolysis (regarding storage, which is one of three primary liver functions—the other two being detoxification and synthesis)
- About £1.4 million over five years
- Nine principal investigators (PIs) from diverse disciplines (including math, computer science, physiology)
- Six postdocs at present (two each from math, comp sci, and life science)
- One objective of overall BEACON program is transfer to industry
  - But UCL PIs say industry isn't interested because they don't have a product that can be sold
  - Will also add two faculty members from business school, to explore commercial prospects
- Year 1 work settled on base system and began development of software infrastructure
  - Focused on how to integrate modeling across multiple biological scales
    - Gene / protein / cell-cell communication
- Year 2 work is now incorporating life science postdocs to generate appropriate new data
- Philosophy is not to pursue large-scale data generation exercise then analyze by informatics methods (i.e., not "data-driven," "bottom up"); instead, approach is "systematic simplification," starting with physiology and working down to underlying mechanisms at successive levels ("model centric," "top down").
- Emphasis on developing software schema for handling models at different scales

    - Some connection to overall U.K. computer science Grand Challenge topic of "*In Vitro* to *In Silico*"
- Team PIs have experience in other systems
    - Hearing
    - Inflammatory system cytokine/cell interactions

Perspective

- Rate-limiting issue in systems biology is more "data integration" than "computation"
- Modeling approach is heterogeneous
    - Some differential equations
    - Some process calculus
    - Some state-space analysis
- Focus on a set of pathways, and study it from gene level to protein level to cell function level
- Goal is to "model," instead of to "simulate"
- Contrast to some other BEACON projects, which are more "genome-wide, full-detail, informatics" oriented rather than "physiology-aimed, model-centric"
- Important to address storage of information associated with experimental data, such as protocols used
    - Currently available data-base software is deemed unsatisfactory for this purpose
    - This issue is exacerbated by contemporary publishing practice, in which methods details are streamlined
- Systems Biology = Modern Physiology
- "Cheap wins" ("low-hanging fruit?") are considered to be education and supporting tissue engineering efforts at UCL and elsewhere (including industry)
- In BEACON liver project, aim is to make some specific, non-intuitive predictions about pathway operation
- Current pay-offs are coming more from "small-scale models;" can this kind of work be termed "systems biology?"
- Middle-ware of physical integration of different models is an easily solved problem
    - More difficult problem is "biological compatibility"

Site: **Université Libre de Bruxelles**
Faculté des Sciences, Service of Physical Chemistry
Building NO, Boulevard du Triomphe
B-1050 Brussels, Belgium

Date visited: July 8, 2004

WTEC Attendees: M. Cassman (Report author), A. Arkin, F. Doyle, F. Heineken, H. Ali

Hosts: Albert Goldbeter, Professor, Unit of Theoretical Chronobiology, Tel: +32-2-650-5772, Email: agoldbet@ulb.ac.be

BACKGROUND

The Free University of Brussels (the French part, ULB) has about 20,000 students, with the Faculty of Science including a variety of disciplines, but not medicine, which has a separate Faculty. There is also a Faculty of Engineering, with several groups interested in biological systems. The university has a long tradition of theoretical biology, initiated by eminent scientists such as the late Ilya Prigogine (1917–2003), winner of the 1977 Nobel Prize in chemistry, who was on the faculty. Dr. Goldbeter is in the Faculty of Science, in the chemistry section, where most of the theoretical biology is located. The focus of his group is the modeling of dynamic biological phenomena, particularly circadian rhythms and metabolic oscillations. His section, within Physical Chemistry, is the Unit of Theoretical Chronobiology, and has three senior members: besides himself, this Unit includes G. Dupont (Ca++ signaling) and J.C. Leloup (circadian rhythms). Other scientists involved in theoretical biology, within the service of physical chemistry, include M. Kaufman (dynamics of regulatory gene networks and theoretical immunology); R. Thomas (logical analysis of regulatory circuits in gene networks); R. Lefever (patterns of vegetation and theoretical ecology); J.C. Deneubourg (theoretical and experimental study of social insects). Also involved in theoretical biology are T. Erneux from physics (bifurcation analysis, neurobiology) and S. Swillens from medicine (Ca++ signaling). Additionally, similar modeling efforts exist in the departments of oceanography, theoretical ecology, and neurobiology. The ULB also has a research group in bioinformatics within the Faculty of Science (S. Wodak, J. Van Helden). All in all, because of this longstanding tradition, the position of theoretical biology at the ULB is a relatively strong one on the European scene. Although Dr. Goldbeter is not himself an experimentalist, he has engaged in numerous collaborations with experimentalists both at the University and elsewhere in Europe or the U.S. The research is not focused around large collaborative efforts, but is largely performed in small groups. Many of the above-mentioned groups are nevertheless part of Networks of Excellence or Integrated Research Projects organized by the European Union.

There appears to be limited opportunity for expanding the research effort in the University. Full-time positions are hard to come by. (These are often filled through the Fond National de la Recherche (FNRS), an NSF-equivalent. Two such tenured positions were recently provided to investigators in Dr. Goldbeter's group.) The University has recently appointed a young professor (J. Van Helden) in bioinformatics, who joined the group of S. Wodak. Another professor recently appointed in the informatics department is G. Bontempi, who also works in bioinformatics. Few funds are available for postdocs from grants, except from the European Commission. Graduate students have other sources of support, including the possibility of funds from the FNRS or from other governmental institutions.

The University is actively trying to adapt to the European Bologna agreement of 2002 which attempted to synchronize higher education across Europe. The intent is to make it easier for students to move from one university to another. The courses have standardized credit assignments to facilitate this. The agreement provides for three years for the BS and two years for the MS (i.e. a total of five years, instead of a total of four years currently required for obtaining the final "Licence" diploma), and three years for the PhD (The PhD portion does not seem to be as well defined, based on other comments heard by the WTEC team during the trip.) In the Faculty of Science to which Dr. Goldbeter belongs, an MS degree in bioinformatics and modeling is being established. This would have three orientations, one of which would be chosen by the students during the second year of the MS: "Classical" bioinformatics; computational structural biology; and modeling of dynamic biological phenomena. Students will come from anywhere in the faculties of science and engineering. However, they may or may not have training in biology or mathematics/informatics prior to the MS. During the first year of the MS, courses are designed to remedy this. Students are required to take broad courses in each of the three orientations. The MS thesis would require four to six months of lab work. Like the other MS programs, the program in bioinformatics and modeling is scheduled to start in the academic year 2007–2008.

Site: **University of Warwick**
Mathematics Institute
Coventry CV4 7AL, U.K.

Date visited: July 7, 2004

WTEC Attendees: F. Katagiri (Report author), D. Lauffenburger, C. Stokes

Hosts: Andrew Millar, Department of Biological Sciences,
Tel: +44 (0)24 7652 4592,
Fax: +44 (0)24 7652 3701,
Email: Andrew.Millar@warwick.ac.uk
Nigel Burroughs, Department of Mathematics,
Tel: (0)24 7652 4682,
Fax: +44 (0)24 7652 4182,
Email: njb@maths.warwick.ac.uk
Jim Beynon, Warwick HRI, Tel: +44 (0)24765 75141,
Fax: +44 (0)24765 74500,
Email: jim.beynon@warwick.ac.uk
Greg Challis, Department of Chemistry,
Tel: +44 (0)24 7657 4024,
Fax: +44 (0)24 7652 4112,
Email: G.L.Challis@warwick.ac.uk

OVERVIEW

Major programs and training (based on Millar's and Burroughs' presentations):

Interdisciplinary Program in Cellular Regulation

The Interdisciplinary Program in Cellular Regulation (IPCR; http://www.maths.warwick.ac.uk/ipcr/) is a four-year, approximately 1.5 million program funded through a Multidisciplinary Critical Mass Research Activity grant from the Engineering and Physical Sciences Research Council (EPSRC) and Biotechnology and Biological Sciences Research Council (BBSRC). It is directed by David Rand (Dept. of Mathematics) and co-managed by Millar and Burroughs. It involves the mathematics, statistics, and biological sciences departments. The program aim is to train eight postdocs with backgrounds in mathematics so that they are equipped to study biological problems. The program is focused on theoretical analysis, and the funding does not support experimental data generation. Millar and Burroughs got separate grants for experiments to complement IPCR efforts. IPCR has projects organized under two themes:

Theme 1, Networks and data, which is managed by Millar; and Theme 2, Spatial process and regulation, which is managed by Burroughs. The strength of IPCR is that it sets theoretical work for biological problems at many different levels. It includes use of Bayesian inference to build a network model based on different types of large data sets, such as expression profiles, statistical models for parameter estimation for a relatively small signaling network such as a circadian clock, development of software to interface different databases, and protein sorting on the membrane. Postdocs are trained in biology through seminars, journal clubs, and attendance in group meetings of biology labs and one-afternoon symposia designed for the program. They take induction courses for Molecular Organization and Assembly in Cells (MOAC) students (see below). They are also paired with biologists for particular projects.

Molecular Organization and Assembly in Cells

Molecular Organization and Assembly in Cells (http://www2.warwick.ac.uk/fac/sci/moac/) is a PhD program that integrates areas of mathematics, biology, and chemistry. This program is funded by a seven-year Doctoral Training Center grant which started in 2003. The program starts with a six-month induction course for the three areas, followed by a lab rotation in each of the three areas before students choose labs for their PhD projects. Most students entering the program usually have backgrounds in physical chemistry.

The WTEC attendees and hosts discussed the problem of highly specialized higher education (especially at the undergraduate level) in the U.K. It would be more desirable to have broader education to train researchers in interdisciplinary areas, such as systems biology. The anticipated European standards for higher education, which consists of three-year undergraduate, two-year MS, and three-year PhD programs, will likely affect higher education in the U.K., which is more accelerated than the proposed EU standard.

RESEARCH

Millar presented the biological sciences department's efforts towards systems biology. The department has about 55 research groups, the majority of which work on a wide variety of topics in molecular and cell biology. Efforts towards systems biology in the department are represented by (1) identifying key regulatory points as targets for interventions, such as c-myc work by Michael Khan and respiratory tract virus work by Andrew Easton, (2) transmembrane protein transport work by the Robinson/Toxin group, and (3) biological clock work by Millar. Millar uses the model plant *Arabidopsis thaliana* for study of a circadian clock, which is comprised of a small sub-network involving 10–20 genes. His mathematical interest is identification of design principles of the clock that apply across species.

Burroughs discussed what he thinks systems biology should incorporate. The biological problems he is interested in are in the area of immunology and population genetics. They include: (1) Gene flow. Developing a mathematical method to handle phylogenetic incongruency based on insertions and deletions in the genome, detected by DNA arrays. (2) Stochastic modeling of signal cascades, using queueing theory and large-deviation theory. (3) Spatio-temporal sub-cellular systems. He models the distribution of cell-surface proteins in immunological synapses. For example, major histocompatibility complex (MHC) molecules localize in the synapse between NK and B cells. His model, using partial differential equations, predicts that the MHC molecules form discriminating concentric patterns according to the distance defined by the pairing molecules between the membranes of two cells. He has three postdocs working in labs of various collaborators to collect experimental data to determine parameters, such as membrane elasticity, for the model and to test the model by obtaining reconstituted 3D images of the molecule localization.

Beynon and Challis are not directly involved in IPCR.

Beynon presented the generation of large-scale resources for *Arabidopsis* research. CATMA (complete *Arabidopsis* transcript microarray) aims to produce a microarray with Polymerase chain reaction (PCR) products printed. Version 2 covers 28,052 genes. It started as a voluntary consortium of U.K. researchers and later developed into a full-scale EU-funded project. CAGE (Compendium of *Arabidopsis* Gene Expression), led by Martin Kuiper, generates standardized expression profile data and makes them publicly available through ArrayExpress. AGRIKOLA (*Arabidopsis* Genomic RNAi Knock-out Line Analysis), led by Ian Small, systematically knocks out *Arabidopsis* genes using RNAi technology. SAP (systematic analysis of promoters) generates promoter-reporter transgenic lines in a systematic manner. All the resources are publicly available.

The EU-funding mechanism was discussed. It requires multinational collaboration, while maintaining full flexibility for type of institution and for more than one partner per country. It is aimed at creating world-class teams of expertise in Europe, and is creating virtual centers for large-scale biology research, which will help to improve the quality and efficiency of data/resource generation in large-scale efforts.

Challis discussed potential applications of systems biology in studies of biosynthetic networks that are not well known. Actinomycetes, such as *Streptomyces* species, produce complex compounds including many pharmaceutical products, such as antibiotics. The *Streptomyces coelicolor* genome has been sequenced. He is developing liquid chromatography, solid phase extraction, nuclear magnetic resonance and mass spectrometer

(LC-SPE-NMR-MS), which is to facilitate structure determination of unknown compounds. He hopes that the combination of genome information, gene knock-out, and efficient compound structure determination will lead to a better understanding of the biosynthetic networks. Such knowledge may be applied to efficient production of intermediates that can be used for rational synthesis of pharmaceutical compounds and to exploration of metagenomics, which is to create new compounds by combining biosynthetic genes from different organisms.

The hosts and the WTEC team also discussed the situation in modeling research, wherein chemical engineers are leading in the U.S. while mathematicians are leading in the U.K.

Challenge

Not much bioinformatic effort has been devoted to creating efficient ways to handle large data sets. Such an effort will be crucial for incorporating large data sets into modeling projects.

Relationships with Industry

Although they view the pharmaceutical industry and the crop science industry as natural partners, no strong interactions with industry have been established.

Site: **Vrije Universiteit (Free University) Amsterdam**
CRBCS/FALW, De Boelelaan 1097
1081 Amsterdam, Netherlands

Date visited: July 9, 2004

WTEC Attendees: F. Doyle (Report Author), A. Arkin, M. Cassman, F. Heineken, H. Ali

Hosts: Hans Westerhoff, Tel: +31-20-444-7230, Email: hans.westerhoff@falw.vu.nl

BACKGROUND

The host (Hans Westerhoff) is embedded in the BioCentrum Amsterdam, which, with its recent refocusing on molecular systems biology, has become engaged in various collaborative projects with many Dutch groups, and involved as adviser in a number of systems biology programs in Europe. As such he organized a full program consisting of: (i) discussions about international developments in systems biology, (ii) an overview of European developments in systems biology, (iii) Dutch developments in systems biology, (iv) examples of Dutch research projects in systems biology, and (v) a presentation on educational initiatives in Holland in systems biology. The fourth topic consisted of a series of brief scientific presentations by various professors and graduate students from representing various locations in The Netherlands (including Delft, University of Amsterdam, The Wageningen Centre for Food Science and Groningen University).

In the international developments in systems biology, Prof. Westerhoff highlighted a number of institutes and consortia that have emerged in recent years including the Institute for Systems Biology (ISB) in Seattle, the Alliance for Cellular Signaling, various modeling efforts (virtual cell, silicon cell, and e-cell), conferences (namely the International Conference on Systems Biology (ICSB), which will be hosted by Europe this fall [October 9–13]), and several topic-oriented systems biology alliances. The last topic generated significant discussion, as these were grass roots or bottom-up initiatives that have attracted an international following. The International *E. coli* Alliance has already held two conferences, is producing a white paper, and is formulating strategies and recommendations for funding. In that area, Genome Canada now seems to be the funding organization most actively interested in funding this initiative. Prof. Westerhoff mentioned that there was pressure to expand the initiative beyond *E. coli* to other prokaryotes. He expressed concern that *E. coli* was a difficult systems biology topic to get funded because too many groups are involved in some kind of research in that organism without being interested in the functioning of the organism as such. Also, he felt that it was essential that the U.S. play a role in supporting it, as nation where most *E. coli* research of high quality is being done.

The Yeast Systems Biology Network has convened two working meetings, and has organized committees to deal with data acquisition and databases, modeling and simulation, training, and dissemination. Their funding strategy is also under development, but remains unclear, at present. There was brief discussion of a nascent effort for forming an alliance for Growth Factor Signaling, organized by Kholodenko and Goryanin (first meeting now in Japan, January 2005, at RIKEN).

With regard to specific European developments, Prof. Westerhoff reviewed both national programs in Germany, Finland, The Netherlands and the U.K., as well as the transnational programs including Systems Biology of Microorganisms (SysMO) and ERANet. In addition, he reviewed EU initiatives and the EUREKA program. The SysMO program is coming from the German Federal Ministry of Education and Research (BMBF), and is aiming to bring in Austria, Holland and others for transnational efforts in microbial systems biology. SysMO attempts to deal with the problem (perhaps unimaginable to the U.S.) that 95% of funding of European Sciences runs through the governments of the individual EU members states. As systems biology programs tend to require the involvement of groups from different disciplines, a Dutch group working on the systems biology of organism X is now forced to take another Dutch group working on organism X as partner in a grant proposal, whereas a much better group on the topic may be only 100 miles away but across the German border. SysMO's target call for proposals is 2005; funding of scientific research is to start early 2007. A related activity, ERANET, has been catalyzed by Brussels, and has a Paris meeting planned in August. The aim would be to coordinate funding on national programs by involving the national research councils; ERANET will fund the national funding organizations to coordinate; it will not fund actual research. The EU initiatives were largely covered by a separate site visit to the European Commission. The Dutch involvement in the FP6 program included proposing a Network of Excellence on systems biology. Explicit support for systems biology by the European Commission FP6 program has been limited to a so-called Specific Support Action, called EUSYSBIO, which does not fund science but a reconnaissance of what European systems biology might become. As part of this EUSYSBIO, the Dutch run a workshop on standards (to be done at ICSB 2004), help formulate a white paper and/or a textbook on the subject of Systems Biology, and organize a training course on systems biology (http://www.febssysbio.net). The EUREKA program is an industry-driven initiative, whereby funds are committed by industry to targeted programs. The idea is to encourage federal or national support to follow from the various agencies. Given the industry impetus, the focus is on biotech problems in food production and the pharma sector. They have a virtual cell initiative (InSysBio), which was just approved in June 2004. The final item reviewed was the European Science Foundation (ESF), which was a joint effort among national science foundations that aimed to catalyze common research programs. It was reported that systems biology was under evaluation as possible ESF theme. The niche area for European efforts in systems biology was "bottom-up" approaches that aim to understand the special systems properties/principles that yield behavior in complex networked biological systems.

As a summary of the discussion on European initiatives, it was clear that successful efforts must involve both "wet" and "dry" systems biology (experimental and computational efforts). It was also clear that collaborative partnerships were in the forefront of European strategic planning, both within Europe, as well as with the U.S., Canada, Japan, Korea and China.

Research Projects

A number of Dutch initiatives in systems biology were then reviewed, including the NWO (Netherlands Organization for Scientific Research, hybrid between National Science Foudnation (NSF) and National Institutes of Health (NIH)) that was funding projects in bioinformatics, molecule to cell, and computational biology at a level of $6 million each. There are also initiatives in National Genomic Centers (funded at $200 million), including two that address some aspects of systems biology (Center for Medical Systems Biology with Leiden, VU, TNO, and the Kluyver Center). Future efforts were described in the form of a set of focused program proposals (so-called SBNL) that focus on organisms (*L. lactis* and *S. cerevisiae*) as well as tools (the Silicon Cell).

The program continued with short (15 minute) presentations from a number of scientists on focused topics including:

- **Prof. Barbara Bakker** (Free University)—"Vertical Genomics"—Described efforts aimed at understanding functional properties of the network in terms of interaction of components, specifically the role of metabolic effects as opposed to hierarchical regulation. The Bakker research program is part of a six member consortium, each of which provides specific technological capabilities to the group, i.e. Mass Spec, microarrays, enzymatic measurements, etc. They develop common conditions and protocols for the experiments, have retreats, and in general have a great deal of coordination.

- **Jurgen Haanstra** (Free University)—"Network-based Drug Design"—Discussed the problem of inhibiting a vital metabolic flux in a pathogen without harming the host, with implications for targeting. This required a combined mathematical/experimental study of the network in both the parasite and the host. The pathogen under study is a trypanosome, of considerable medical and economic interest.
- **Frank Bruggeman** (Free University)—"Silicon Cell"—Described the initiative to construct computer replicas of parts of living cells from all detailed experimental information. Such 'Silicon Cells' can then be used in analyzing functional consequences that are not directly evident from component data but arise through contextual dependent interactions. Silicon cell is different from the U.S. Virtual Cell (which provides a toolbox for modeling) and Japanese E-Cell (which makes approximate, phenomenological models) in that it is precise, enabling testing of hypotheses on system functioning, in that it is web-based (www.siliconcell.net), and in that it engages a number of scientific journals publising systems biology models also at the reviewing stage.
- **Jorrit Hornberg** (Free University)—"Signal Transduction"—Demonstrated the utility of applying metabolic control analysis (MCA) to signaling pathways to elicit understanding (for example, role of phosphotases versus kinases for amplitude versus duration performance). He showed new laws discovered by systems biology for signal transduction.
- **Wouter van Winden** (Delft University)—"Systems Biology in Delft"—Spoke about the efforts in the Bioprocess Technology Group at Delft, including linear logarithmic kinetic modeling, dynamic metabolite measurements, and future plans for experimental design (BioScope). Dr. van Winden is part of a collaboration team that includes Dr. Bakker.

Summarizing the research presentations, it is worth noting that MCA and its extensions to signal transduction and gene expression (hierarchical control analysis, HCA) play a key role in many of these studies. MCA and HCA relate the control and regulation of intracellular systems to the properties of their components. Prof. Westerhoff is a leading contributor to this 'integrative' approach to systems biology. The Westerhoff group is special in combining successfully experimental and theoretical systems biology in a single group, where the latter activity includes both modeling and the discovery/proving of new laws and principles specific to biological systems.

Prof. Westerhoff concluded the program with a discussion about teaching systems biology in the Netherlands. He described in detail a new MS program in Biomolecular Integration/Systems Biology. The aim was to provide both expertise in advanced conceptual and modeling methodologies as well as insight into important biological/biomedical issues. It is a two-year program, with 120 credits (European Credit Transfer System—ECTS), and involves a detailed research project where the student spends half the time in Amsterdam and half the time with a partner group in a different country, both locations being involved in the advising. An example would be a study of a network in organism X, experimentally in Amsterdam and modelingwise in Berlin (or vice versa). Teaching efforts at the PhD level were still evolving at the time of our visit, and involved a joint graduate school with Humboldt University.

There was an extended discussion of the challenges that one faces in teaching systems biology, including the demanding curriculum, low perception by peer groups, lack of industry demand for systems biology majors, and lack of interest by young students in science. There was also some discussion of the different approaches to creating systems biology "specialists," including the efforts of another university in the Netherlands to create an undergraduate degree in the area, thus generating students that do not require synchronization (from multiple disciplines) at the MS level. The balance of the discussion seemed to favor the interdisciplinary MS route with undergraduate majors coming from traditional departments (physics, math, computer science, biology).

APPENDIX C: SITE REPORTS—JAPAN

Site: **Computational Biology Research Center (CBRC)**
The AIST Tokyo Waterfront
Bio IT Research Building
2-42 Aomi Koto-ku
Tokyo 135-0064, Japan
http://www.cbrc.jp/

Date: December 14, 2004

WTEC Attendees: C. Stokes (Report author), M. Cassman, A. Arkin, F. Katagiri, F. Doyle, S. Demir, R. Horning

Hosts: Yutaka Akiyama, Director, Tel: +81-3-3599-8087, Fax: +81-3-3599-8085, Email: akiyama-yutaka@aist.go.jp
Makiko Suwa, Deputy Director, Tel: +81-3-3599-8051, Fax: +81-3-3599-8085, Email: m-suwa@aist.go.jp

Dr. Akiyama described the CBRC goals and research projects. The center's deputy director, Makiko Suwa, also attended.

BACKGROUND

CBRC is a research center established in 2001 in the National Institute of Advanced Industrial Science and Technology (AIST). It is dedicated to computational biology (bioinformatics) research, with a focus on developing technologies for gene finding, protein structure prediction, protein-protein interactions, gene networks, and mass spectrometry (MS) analysis, and utilizing these for proteomics research and commercial applications. They have a small wet lab facility, including mass spectrometry devices to do research on measurement technology for proteomics. There are 120 total members, including 19 full-time researchers and 16 postdocs. Their basic research becomes applied very quickly given the nature of their work, and Dr. Akiyama estimated that 30–40% of their research might be considered "basic" with the rest "applied." Their measures for success include patents, income from industry, and publishing in good journals.

RESEARCH

CBRC is organized around three areas of research, genome informatics (algorithms, mathematical models, and sequence analysis), molecular function (biomembrane informatics, protein function, and molecular modeling and design) and biological systems (cellular informatics and high-performance computing). Dr. Akiyama noted that, when founded, CBRC planned a systems biology focus, but they felt there was not enough data to support that so instead they started by working to build the elementary technology to support systems biology (proteomics and network analysis).

Gene Finding

For gene finding research Dr. Asai has developed and utilizes their GeneDecoder system (www.genedecoder.org), using hidden Markov models and related approaches. Dr. Suwa and others have found many putative G protein-coupled receptor (GPCR) genes from the human chromosome and made DB-"SEVENS" (http://sevens.cbrc.jp) and are working to predict GPCR-ligand interactions based on sequences (in collaboration with Mitsubishi Chemistry). The approach is fully automated so they are applying it now to several organisms (bacteria, yeast, *Arabidopsis*, *Oryzae*, *Drosophila*, *C. elegans*, human). The approach looks for common structural motifs rather than sequence homologies, which are sometimes in short supply between the organisms. Large-scale hidden Markov algorithms are used to do these sophisticated splicing predictions. Another project focuses on finding genes of Aspergillus oryzae, in collaboration with companies in the fermentation industry. They will start a new venture company based on findings on this organism. The industry paid for the sequencing, while CBRC did the analysis. This work is being prepared for publication.

Protein Structure Prediction

The CBRC uses a profile-profile comparison approach for protein structure prediction. For this, they developed and use the FORTE-SUITE (Tomii, 2004), which can verify 10,000 models simultaneously. At the Critical Assessment of Techniques for Protein Structure Prediction 5 (CASP5) and CASP6 competitions they took 16th and 3rd (of 200) places, respectively, in the fold recognition category. Another focus area is the development of 3D structure models from genome sequences. Dr. Hirokawa is also working on docking of ligands with proteins, an area that also requires experimental work, with actual applications for virtual screening of candidate drugs using the predicted protein structures. In one project, he has predicted that new compounds found through such a screen will act as histamine receptor antagonists and are putative new drug leads.

Gene Network, Protein-protein Interaction

Time-course genomic data is very sparse and they found they needed new technology to estimate gene regulatory networks. Another team at AIST (Dr. Miyake) built transfection arrays with 8,000 spots to do systematic experiments for ribonucleic acid interference (RNAi) and complementary deoxyribonucleic acid (cDNA) transfection to living cells. They utilize this in conjunction with an imaging system for real-time tracking for each cell and observe them for 24 hours. One project involves observing cell growth as a function of RNAi and cDNA transfection. Dr. Horton and colleagues gather the time series and DNA chip data in their "Cell-Montage" database (not yet public) with the purpose of categorizing the data automatically to discover if the chip results are related to a pathology. One focus is to identify genes associated with lung cancer.

MS Analysis

For large-scale proteomics use, a major problem using MS/MS information processing is speed of computation. Dr. Akiyama has developed and uses the "CoCoozo" system (http://www.cbrc.jp/cocoozo) for high-speed intelligent MS/MS spectrum analysis. This system is 30x faster than single processor versions of commercial software. A free trial is available on the CBRC web site. One application area is the structure of glycoproteins. Dr. Takahashi is working on a new fragmentation technique that uses a frequency-controlled infra-red pulselaser to first cut peptide bonds and then the oligosaccharide bonds in order to find out where the sugar is attached on the protein. Dr. Akiyama noted that glycosaccharide research a strong point for Japan. This is funded by New Energy and Industrial Technology Development Organization (NEDO).

BUDGET FOR RESEARCH

The CBRC budget is about $8 million/year for research presently, including about $3 million/year from the Millennium project, $2 million/year from industry, and $3 million/year from AIST (about $0.5–1 million/year as base funding and competitive awards forming another $2–2.5 million/year). The industry funding is very important because the government is likely to provide more funding because of it.

TRAINING AND TRAINING BUDGET

The CBRC budget includes another approximately $1.2 million/year from Ministry of Education, Culture, Sports, Science and Technology (MEXT) for training; about 80% is used for "on the job" training, lectures, and seminars for about 20 postdocs and about 35 trainees, and about 20%

pays for annotator training at BIRC. CBRC runs some training courses for their own staff as well as researchers from other organizations. The post-docs typically come from math, biology, and physics and have two- to three-year contracts for their training in the bioinformatics and computational biology area. They can apply for full-time AIST tenured jobs at the end of a couple of postdocs, although those positions are difficult to obtain.

COMPUTING RESOURCES

To support these computationally intensive activities, CBRC has significant supercomputing resources, with a 14.4 Tflop/s PC cluster system and an IBM Blue Gene/L system (22.7 Tflop/s) set in place since February 2005. These facilities are used for protein structure prediction, molecular docking and gene finding research.

COLLABORATIONS

CBRC researchers can freely collaborate with outside labs, and have tight collaborations with industry and universities. Outside CBRC collaborations include more than 10 universities and 10 industry contracts, including the University of Tokyo, Nara Institute of Science and Technology (NAIST), Waseda University, Keio University, Tokyo Institute of Technology, Tokyo University of Agriculture and Technology, Mitsubishi Chemical, Sumitomo Pharmaceuticals Co., Ltd., Kawasaki Heavy Industries, Ltd., NEC, Japan Electron Optics Laboratory Co., Ltd., TSUMURA & Co., Intec Web and Genome Co. Ltd., and Fujitsu FIP.

The collaborations are primarily proteomics- and computer science-oriented. The nature of the collaborations with industry varies; sometimes a company will have researchers work at CBRC, and other times the company will supply materials, or the collaboration is purely scientific. CBRC and the company share in resulting patents and intellectual property. Sometimes CBRC will train programmers for a company as part of collaboration. Professor Akiyama is currently engaged in setting up collaborations within Asia, including Korea, Singapore, China and India.

REFERENCE

Tomii, K. and Akiyama, Y. 2004. FORTE: a profile-profile comparison tool for protein fold recognition. *Bioinformatics* 20: 594–595.

Site: **Japan Biological Information Research Center**
AIST Tokyo Waterfront
2-41-6 Aomi, Koto-ku
Tokyo 135-0064, Japan
http://www.jbirc.aist.go.jp

Date: December 14, 2004

WTEC Attendees: A. Arkin (Report author), M. Cassman, F. Doyle, F. Katagiri, C. Stokes, R. Horning, S. Demir

Hosts: Kimitsuna Watanabe, Director, Tel: +81-3-3599-8101, Fax: +81-3-5530-2064, Email: kwatanab@jbirc.aist.go.jp
Nobuo Nomura, Deputy Director, Tel: +81-3-3599-8137, Fax: +81-3-3599-8141, Email: nnomura@jbirc.aist.go.jp
Tadashi Imanishi, Team Leader, Integrated Database Team, Tel: +81-3-5531-8550, Fax: +81-3-5531-8851, Email: imanishi@jbirc.aist.go.jp

OVERVIEW

Kimitsuna Watanabe, the director of the Japan Biological Information Research Center (JBIRC), was not available the day the WTEC team arrived at the center. However, he was ably represented by functional genomics group leader Nobuo Nomura. Until recently, Prof. Watanabe was also associated with the Department of Integrated Biosciences, Graduate School of Frontier Sciences, University of Tokyo. He is a leader in the field of ribonucleic acid (RNA) biology, in particular, though not limited to, tRNA biochemistry, function and structure. He also leads the structural genomics group. Nobuo Nomura is a deputy director of the Institute as well as group leader in functional genomics and the sub groups of protein expression and cellular function. He has been instrumental in the construction of the full-length long complementary deoxyribonucleic acid (cDNA) collection of Japan (FLJ) project and in using this resource to drive structural and functional genomic studies.

HISTORY

JBIRC has been operating for over three years now and seems to be in full swing. The Ministry of Economy, Trade and Industry (METI) provides the main support for the center through two routes. The first is through indirect support via the National Institute of Advanced Industrial Science and Technology (AIST), a quasi-governmental organization that houses about 2,500 internal researchers, and support of a non-research agency called New Energy and Industrial Technology Development Organization (NEDO), a funding agency established by the Japanese government in 1980 to develop new oil-alternative energy technologies and which has since expanded its mission. NEDO is both a science-funding agency and governmental policy consultant. Like NSF, academic rotators take on program management responsibilities.

METI funds, directly, or indirectly through NEDO, a consortium called called the Japan Biological Informatics Consortium (JBIC, http:// www.jbic.or.jp/bio/english/index.html.) JBIC was founded in 1998 (pre-Millennium project) and reorganized in 2000 as a non-profit organization overseen by a number of ministries with an interest in biotechnology. The organization's mission is to streamline research and development in the biotechnology industry and the creation of new industry through application of informatics technology. JBIC is a consortium of over 90 industry partners which support academic institutions, industrial companies (both for development and services like sequencing) and, ultimately, JBIRC.

While this industrial tie-in does generate some intellectual property (IP) issues among the government and industrial funding, there is evidently an informal and navigable process to decide which IP goes to AIST and which goes to JBIC up front. JBIRC is probably the major beneficiary of JBIC. JBIRC is supported by both JBIC and AIST. This funding is approximately ¥2,332,000,000 per year (about $22,612,236). This funds 71 researchers in structural genomics, 48 in functional genomics, 61 in the integrated database group and 24 staff members. Most members of the Institute are full-time though a few (especially in the structural group) hold concurrent appointments at a university or sites like the National Institute of Genetics (http://www.nig.ac.jp/index-e.html).

JBIRC opened in April, 2001, in the Tokyo Bay Area and Tsukuba as one of the research centers of AIST under the control of the Ministry of Economics, Trade and Industry (METI). The central mission of JBIRC is to exploit post-genomic research. The Institute has three major sections: functional genomics, structural genomics and genome databases. There is a decided focus on research that leads directly to medical and industrial application. As such they aim to develop new technologies to more precisely measure biomolecules and their concentration. Seventy to eighty percent of the Institute is dedicated to wet biology and the rest to informatics. JBIRC is considered to be a unique organization in Japan because it is neither a national laboratory like RIKEN nor a particular arm of a university but it nonetheless is a large-scale operation. The building and associated infrastructure is absolutely top-quality.

Research emphases are on the structure determination of membrane proteins, functional analyses of human full-length cDNAs and the construction of an integrated genome database. The center performs basic research that may lead to medical and industrial uses. JBIRC has 70–80% wet research, with the rest informatics.

STRUCTURAL GENOMICS GROUP

Dr. Watanabe leads the structural genomics group. They are particularly focused on developing new techniques for protein purification and structure determination, in particular for membrane proteins. In addition, they aim to develop analyses and tools for predicting and designing protein/ligand interactions. They are developing an effort in the accurate computational modeling of membrane proteins and *in silico* screening of ligand-receptor binding. The group consists of four subteams in structural analysis, structural informatics, molecular recognition and molecular functional analysis. Their main experimental technologies are nuclear magnetic resonance (NMR), X-ray crystallography and Cryo-EM.

Targets for the group are chosen in two ways. First, they are chosen for academic reasons: the impact of the structure, the feasibility for crystallization/purification. Second, they are chosen for industrial reasons such as medical or industrial importance.

FUNCTIONAL GENOMICS GROUP

This group, lead by Dr. Nomura, bases a good deal of their work around a large collaborative Japanese project to determine the sequence of 30,000 full-length cDNA from the human genome (FLJ). They have been converting these clones into the Gateway system for protein expression (18,000 already done). They've produced two types, normal and fusion ones. There are 40,000 Gateway clones of these different types derived from the original 18,000. This resource supports both the structural genomics and functional genomics efforts.

They use both microarrays and a version of quantitative reverse transcription-polymerase chain reaction (RT-PCR) to follow transcription in human cells. Right now they track about 20,000 cDNAs. The protein network team is dedicated to the experimental identification of functional protein complexes. They generally use mass-spectrometry methods for doing this and claim to be able to isolate these complexes from much smaller samples than are generally needed by other methods. Finally, the cellular function team has a diverse set of approaches to perturbing and measuring cell function including methods of automating gene transformation, scoring the morphology of cells, and determining protein localization in human cells. For this latter effort, 50% of the clones they have tagged localize well and reproducibly with both N- and c-terminally tagged proteins. The rest of the clones are being reanalyzed. They may also start a project for antibody production to get multicolor images. Another very interesting technological development was the successful demonstration of a protein chip capable of following kinase activity fairly accurately. The scale and quality of the results was far better than any I have seen before. It was unclear, however, exactly which scientific problems they were approaching with all these technologies though I suspect it was similar to Kodama's project to discover useful disease biomarkers and druggable protein targets. Dr. Nomura mentioned that Dr. Yuji Kohara is trying a similar sort of project with *C. elegans* and Dr. Takahashi Ito is doing complex determination with yeast.

INTEGRATED DATABASE GROUP

The Integrated Database Group is in a separate building from the other efforts and is largely a sequence informatics shop. All the other analyses are done at the site of data generation. Dr. Takashi Gojobori is a deputy director of the JBIRC as well as professor and director of the Center for Information Biology and DNA Data Bank of Japan (CIB-DDBJ) at the National Institute of Genetics. Prof. Gojobori comes to the Center about once a week. Tadashi Imanishi (group leader for the integrated database team) presented this discussion. There is another group on genome diversity in the integrated database group that wasn't discussed.

The main focus of this group seems to be centralized around the production of an integrated human genome annotation web site (http://www.h-invitational.jp/). The project has eight postdocs, 10 rotators, and 30 systems engineers. The database contains information on 41,118 full-length cDNA clones including gene structures, functions, domains, expression (in some cases), diversity, and evolution.

To compile this information c44 research institutes worldwide collaborated, organized through JBIRC and CIB-DDBJ efforts. The emphasis was on strong human curation while leveraging information from six other database sites including the Munich Information Center for Protein Sequences (MIPS), Mammalian Gene Collection at the National Institutes of Health (MGC@NIH), Human Unidentified Gene-Encoded (HUGE), Human Novel Transcripts (HUNT) at the Helix Institute, Chinese National Human Genome Center (CHGC), and FLJ/DDBJ. After the initial human genome sequence and compilation from these sites the data was frozen for curation and annotation. The data has been frozen since July 15, 2002; however, they are in the process of updating with the National Center for Biotechnology Information's (NCBI) latest genome assembly (build 30) along with the European Bioinformatic Institute (EBI) non-redundant

SwissProt/TrEMBL protein set. Computational and human curation is ongoing.

The annotation of genes is done in two steps: first by computational prediction and then human curation. During the computational steps full-length cDNAs are mapped to the genome, open reading frame (ORF) predictions are made and algorithms for sequence similarity and functional motif prediction accomplish functional/annotation predictions. Also structural predictions are made to aid in the annotation. The second step, human annotation occurs continually, but there are large "jamboree" format annotation sessions. For example, the jamboree that occurred just after the data freeze from August 23–September 3, 2002, was attended by more than 118 people from 40 organizations. In that session 20,000 genes were hand-annotated and most of the rest were done by the end of 2003.

A reanalysis and comparison to the new NCBI assemblies was published in PLOS (Imanishi et al. 2004) and can be found at http://www.plosbiology.org/plosonline/?request=getdocument&doi=10.1371/journal.pbio.0020162) with 158 authors. Interesting findings from that analysis include: 41,118 confirmed cDNAs corresponded to 21,037 gene models; 5,155 of these models were unique to H-invDB. In addition, 4% of the human genome sequence is incomplete (with build 30) though it is less than 1% now. Alternative splicing was found to be highly prevalent: they found around 8,553 isoforms in about 300 genes. Standardized human curation classified 19,574 proteins into five classes based on sequence similarity and structural information: 9,139 proteins had functional definitions; 2,503 domain ids and 7,800 hypotheticals. Non-protein coding genes accounted for 6.5% (1377 loci) of cDNAs. Of these, 296 were classified as non-coding RNAs. This was further supported by conformation and strong predicted secondary structure. Expression experiments showed that 28/47 were found expressed in human tissue. Along the way they also identified a large number of Single nucleotide polymorphisms (SNPs).

This database was funded, in part, as part of the millennium project of Japan that will end next March. (An outline of the Millennium Genome Project can be found at: http://www.ncc.go.jp/en/nccri/genome/mgp.pdf). While this project will continue to maintain this database it is mostly through METI, which currently funds the project at ¥800,000,000 per year (about $7,757,199), which will probably drop in the next few years to ¥500,000,000 or about $4,848,249. This funding supports about 55 people.

Overall, JBIRC presented an astounding infrastructure for the measurement and analysis of human biomolecules.

REFERENCE

Imanishi, T. et al. 2004. Integrative Annotation of 21,037 Human Genes Validated by Full-Length cDNA Clones. *PLoS* 2 6: 856–875.

Site: **Kazusa DNA Research Institute (KDRI)**
2-6-7 Kazusa-kamatari
Kisarazu, Chiba 292-0818, Japan
http://www.kazusa.or.jp/

Date: December 15, 2004

WTEC Attendees: F. Katagiri (Report author), M. Cassman, R. Horning

Hosts: Daisuke Shibata, Lab Leader, Fax: +81-438-52-3948, Email: shibata@kazusa.or.jp
Hideyuki Suzuki, Shibata Lab, Fax: +81-438-52-3948, Email: hsuzuki@kazusa.or.jp
Nozomu Sakurai, Shibata Lab, Fax: +81-438-52-3948, Email: sakurai@kazusa.or.jp
Shigehiko Kanaya, Professor, Graduate School of Information Science, Nara Institute of Science and Technology, Tel: +81-743-72-5952, Fax: +81-743-72-5953, Email: skanya@gtc.naist.jp
Masami Hirai, Saito Lab, Chiba University

Overviews of the Institute and research were presented by Shibata.

OVERVIEW

Kazusa DNA Research Institute (KDRI) was established in 1991 with funding from Chiba Prefecture, and the current building at Kazusa Academia Park opened in 1994. Currently, it has seven principal investigators (PIs), including Director Michio Ohishi (four for human research and three for plant research). Everything at KDRI except for Shibata's New Energy and Industrial Technology Development Organization (NEDO) project described below is funded by Chiba Prefecture.

Plant Gene Laboratory 1, which is led by Satoshi Tabata, focuses on genome sequencing. The genomes sequenced by Tabata's group include cyanobacterium, symbiotic bacterium and approximately 20% of *Arabidopsis thaliana* as part of an international collaboration. The genome of the model legume plant *Lotus japonica* is currently being sequenced by Tabata's group.

Plant Gene Laboratory 2, which is led by Shibata, focuses on functional genomics. His group includes about 40 people, including three PhD researchers (permanently at KDRI), six postdocs (one supported by the prefecture), eight technicians (four with master's degrees), one graduate student (Tohoku University, where Shibata has an adjunct appointment), and part-time workers. Shibata's group works on two projects, a plant metabolomics project and a large-scale functional genomics project. The goal of the functional genomics project is better functional annotation of genes involved in secondary metabolism in *Arabidopsis* and *Lotus*. For example, it is not clear which of many genes annotated as hydrolases actually catalyze a particular hydrolytic reaction in a particular metabolic network. In this project, candidate genes for particular reactions or regulatory activities are identified, and the predicted functions of the candidate genes are validated. This project is funded by NEDO from 2001 to 2009 at about $2 million/year. Progress of the project will be assessed at the midpoint of the funding period before the rest of the funding is awarded. Shibata himself will consider the project successful if the functions of 10% of the genes subjected to analysis are determined. This NEDO project was proposed as a collaboration among Shibata's laboratory, two paper companies (Ohji, Nippon), a tire company (Bridgestone), and a company working on plant secondary metabolism (Tokiwa Chemical). These companies also receive funding from the NEDO project. Shibata's group focuses on basic research and provides information to collaborating companies for translational research.

Currently, research in *Arabidopsis* is more advanced than in *Lotus*. More than 4,000 genes out of total about 27,000 genes in *Arabidopsis* are currently annotated as metabolism-related genes. However, the precise functions of most of the 4,000 genes are not known. Based on transcriptome and metabolome analyses, candidate genes for particular reactions or for regulation of particular metabolic pathways are identified, as detailed in Kanaya's and Hirai's presentations below. The candidate genes are over- or under-expressed in *Arabidopsis* T87 suspension culture cells. Use of a suspension culture cell allows them to have little variation in plant cell conditions compared to the use of an actual plant, and consequently reduces biological variation among samples. They have developed an efficient T87 cell transformation method using Agrobacterium. They have also developed a cryo-preservation method for easy storage and recovery of transformed T87 cells lines. Overexpression lines for more than 500 genes have already been generated. Ten independent transformants for each construct are examined for phenotype.

Plants have many metabolites: typically more than 5,000 metabolites in a species, compared to 2,000–3,000 in humans. Therefore, technology for separation of metabolites is crucial in metabolomics of plants. Shibata's group has a gas chromatography time-of-flight mass spectrometer (GC/TOF-MS), a liquid chromatography photo-diode-array-detection mass spectrometer (LC/PDA-MS), and two capillary electrophoresis mass spectrometers (CE-MS—one for cationic compounds and the other for anionic compounds). They will purchase a liquid chromatography Fourier transformation mass spectrometer (LC-FT-MS) and a liquid chromatography time-of-flight mass spectrometer (LC-TOF-MS) next year with funds from the Prefecture. Currently, they collaborate with Dr. Ohta at Osaka University for Fourier transformation mass spectrometry (FT-MS). The high precision of mass measurement by FT-MS (to 0.00001 MW) is very important in compound identification, as illustrated in Kanaya's presentation below. The MS peak data are converted into a table format. The data from transcriptomes (using Agilent 22k array), metabolomes, and phenomes are integrated. These data comprise their view of systems biology.

A metabolic pathway viewer KaPPA-VIEW (Kazusa Plant Pathway Viewer) was developed. It uses the scalable vector graphics (SVG) format. The viewer and 140 pathway maps for the viewer will be available to the public. A feature of the viewer is that for each reaction in the displayed map, the expression of candidate genes for the reaction under certain conditions is indicated. In this way, it is easy to point out candidate genes based on transcriptome data.

The group has many collaborations defined by classes of compounds. For example, they are collaborating with Dr. Nishiya at Tohoku University for study of cell wall-related compounds.

Kanaya presented bioinformatics work for the NEDO project. The first topic was how to convert MS spectra into a tabulated data format. Different MS generate different types of data. For example, an MS combined with chromatography generates data described by the retention time and the m/z, while an FT-MS generates data described by m/z and the peak intensity. Another issue is how to handle small experiment-to-experiment variations in peak positions. Separate modules are used to handle different types of data. After conversion into a table, significant peaks are selected using a t-test, and the selected data are visualized using principal component analysis (PCA). The second topic was estimation of molecular formula from molecular weight data. Increased precision in mass measurement results in reduction of the possible molecular formulae. This illustrates the importance of precise mass measurement by FT-MS. The third topic was a database for candidate compound identification. The database is being populated with about 20,000 microbial compounds and approximately 100,000 plant compounds which have been reported, so that the database holds the compound information and information about species in which a particular compound has been detected. Data for 15,000 compounds have been entered into the database. When the count of compounds was plotted against the number of species assigned to a compound in log scales in both axes, it showed a power law.

Hirai presented examples of candidate identification in *Arabidopsis* by integration of transcriptomes and metabolomes. The first example was sulfur metabolism. Transcriptomes (21.5k genes) and metabolomes (about

2,000 compounds) from *Arabidopsis* treated with +/- sulfate for various lengths of time were collected. Transcriptome and metabolome data were simply combined into a single table and visualized using batch-learning self-organizing map (BL-SOM), developed by Kanaya. In this way, good correlations in the position in the BL-SOM lattice among related genes and metabolites were observed. In the second example, almost all the enzyme genes involved in anthocyanin biosynthesis were shown to be co-regulated. Hirai mentioned that good correlations observed in the examples are not typical among enzymes of primary metabolism. This signifies the advantage of focusing on secondary metabolism in the NEDO project.

Site: **Keio University, Institute for Advanced Biosciences**
Tsuruoka Campus
Baba-cho 14-1
Tsuruoka City, Yamagata, Japan
http://www.iab.keio.ac.jp

Date: December 17, 2004

WTEC Attendees: F. Katagiri (Report author), M. Cassman, A. Arkin, R. Horning

Hosts: Masaru Tomita, Director and Professor, Fax: +81-466-47-5099, Email: mt@sfc.keio.ac.jp,
Hirotada Mori, IAB, Tel: +81-235-29-0521,
Fax: +81-235-29-0529,
Email: hmori@sfc.keio.ac.jp,
Nara Institute of Science and Technology,
Tel: +81-743-72-5660, Fax: +81-743-72-5669,
Email: hmori@gtc.aist-nara.ac.jp

OVERVIEW

Keio University is the oldest university in Japan, founded by Yukichi Fukuzawa in 1858. Now it has five campuses and 22 departments with 28,000 undergraduates, 2,400 master course graduate students, and 950 doctor course graduate students. The Institute for Advanced Biosciences (IAB) was established in 2001 with a funding of $110 million from Yamagata Prefecture and Tsuruoka City. The operation cost of the Institute is about $9 million per year and is covered by funds from the local governments and competitive grants.

Tomita started the E-Cell project in 1995. E-Cell is a software environment for whole-cell simulation consisting of 127 genes, 4,268 molecular species, and 495 reactions. It was initially used for simulation of *Mycoplasma genitalium* in 1997 and has since then applied to many different cells, including the erythrocyte, *E. coli*, and rice. During the initial project it became clear that the bottleneck in generating accurate simulations is the absence of quantitative data. Therefore, when Tomita started IAB, the Institute aimed at interactions among efforts in biosimulation, genome engineering, metabolic engineering, and analytical chemistry, and approximately 80% of the efforts in the Institute are spent on wet lab-based research.

A key technology at the Institute is capillary electriphoresis mass spectrometry (CE-MS), developed by Soga et al. and commercialized by

Agilent. CE gives a resolution comparable to that of gas chromatography (GC) and does not require derivatization. CE-MS reproducibly detects about 800 charged metabolites under a single experimental condition in 30 minutes. Liquid chromatography mass spectrometry (LC-MS) is used for the measurement of about 150 metabolites, primarily lipids. In order to further identify peaks, 1,490 standard compounds were analyzed. (Another 696 standards will soon be added.) Since not all peak identification can be accomplished by use of standards, mass spectrometry approaches for molecular weight and structure determination are also applied. Furthermore, CE migration time of a given compound can be predicted by a neural network approach developed by Sugimoto et al., and also used to identify peaks.

The facility includes:

- 19 x CE systems
- 6 x LC systems
- 2 x GC/MS systems
- 9 x quadrupole MS systems
- 4 x ion-trap MS systems
- 2 x TripleQ MS/MS systems
- 6 x electrospray ionization (ESI)-time of flight (TOF-MS) systems
- 1 x quadrupole (Q)-TOFMS system
- 1 x nuclear magnetic resonance (NMR)

A project to simulate central metabolism in *E. coli* is funded by New Energy and Industrial Technology Development Organization (NEDO). During steady-state growth, ^{13}C compounds are added to the medium, samples are collected by an autosampling machine at various time points, and the labeling state of metabolites are measured. In this way, the flux rates of particular reactions can be estimated. Transcriptomes, proteomes, and metabolomes are also analyzed for the same samples. They expect to compare 200 strains of *E. coli* using these approaches. These analytical methods are also to be applied to many open reading frame (ORF) deletion mutants generated by Tomoya Baba and Mori (see below).

Mori presented his research on the functional genomics of *E. coli*. This was initiated eight years ago at the Nara Institute of Science and Technology (NAIST) with funding from the Japan Science and Technology (JST) agency at approximately $1 million per year for five years. It developed into a collaboration between NAIST and IAB. For cloned gene resources, all the *E. coli* ORFs were cloned with T7 promoters, 6xHis tags, and with or without GFP. For mutant lines, all the ORFs were replaced with the Km^r gene or deleted in-frame (303 genes were found to be essential). These resources are available to the research community from Mori (http://ecoli.aist-nara.ac.jp). This distribution activity is supported by the Ministry of Education, Culture, Science, Sports, and Technology (MEXT). To study essential genes, which are lethal when deleted, such genes were introduced with the isopropyl beta-D-thiogalactopyranoside (IPTG)-inducible promoter on a mini-F plasmid while the endogenous genes were deleted. Through complementation on IPTG expression more than 24 essential ORFs can be deleted in this way. All the chromosomal regions between essential genes were subjected to deletions. Only deletions of a total length of more than 100 kb could not be recovered, probably due to synthetic lethality. Deletion mutants for transcription factor genes were analyzed to identify their transcriptome in a steady state. The indirect effects in transcriptional regulation were, however, too complex to clearly define the interactions. Time course data are now being collected together with software which predicts indirect effects in signaling pathways. With green fluorescent protein (GFP) fusions, subcellular localizations of the proteins were also determined. Using His-tag affinity purification, (successful with 2,700 ORFs, but no membrane proteins), 11,531 protein-protein interactions were detected. The above web site also includes these characterization results.

During a tour of the Institute, a brief overview of some other individual projects was given. Mitsuhiro Itaya's group developed technology to cut and paste large deoxyribonucleic acid (DNA) fragments. This technology will be used to construct a single plasmid construct with many unlinked genes of interest (e.g., genes involved in a single metabolic pathway). Hiroshi Yanagawa's group developed a novel protein-protein or protein-DNA interaction detection technology, in which proteins are physically linked to their own messenger ribonucleic acid interference (mRNA) *in vivo*, so that the RNA sequence can be used as specific tags for the proteins.

Keio University has a master's course graduate program (Bioinformatics) to train students in the interdisciplinary area between biology and computer science (http://www.ttck.keio.ac.jp/Bioinformatics-program/). The curriculum includes computer courses, experimentation courses, and self-study courses in biology.

IAB has a student lab for undergraduate and graduate training. Informatics students are trained in experimental biology for half a year to a full year. Twenty students, who are supported by the local governments, including free dormitory rooms, are trained at a time. Graduate students are given opportunities to extend their stay at IAB and pursue their own projects.

IAB has created a venture company, Human Metabolome Technologies (HMT), 10% of which is owned by Keio University. Many employees of

HMT also have appointments with IAB, and the lab space for the company is located in the same building as IAB.

The campus will be doubled in space in 2006 with funding from the local government. This will give opportunities to expand IAB and HMT and to add incubator space for start-ups. A "village" for 200 families will also be built adjacent to the campus.

Site: **Kitano Symbiotic Systems Project**
6-31-15 Jingumae, M-31 Suite 6A
Shibuyaku, Tokyo 150-0001, Japan
Tel: +81-3-5468-1661
http://www.symbio.jst.go.jp/symbio/sbgE.htm

Date: December 13, 2004

WTEC Attendees: A. Arkin (Report author), M. Cassman, F. Doyle,
F. Katagiri, C. Stokes, S. Demir, R. Horning

Hosts: Hiroaki Kitano, Director,
Email: kitano@symbio.jst.go.jp
Douglas Murray, Researcher, Tel: +81-3-5363-3078,
Fax: +81-5363-3079,
Email: dougie@symbio.jst.go.jp
Noriko Hiroi, Researcher,
Akira Funahashi, Researcher,
Hisao Moriya

OVERVIEW

The Symbiotic Systems Project (SSP) is led by Hiroaki Kitano, who is also the director of Sony Computer Science Laboratories, Inc. (SCSL) and a visiting professor at Keio University. The basic goal of the project is to develop and apply new technology and computational tools to understand dynamical phenomenon in cellular systems. The primary location for the program is housed in a well-designed space in the Harajuku section of Shibuya district in Tokyo. There is also wet lab space 13 minutes away by foot at Keio Medical School. Research areas of interest to the Institute include new approaches to understanding biological robustness, and how software and computational tools can help in addressing such issues. They have a central interest in mechanisms that make cancer difficult to treat and in the control of the cell cycle.

The meeting was held at a well-designed office and computational facility in Shinonamachi. Design is another of Dr. Kitano's talents. For example, illuminated walls of glass that allow researchers to project slides or write on the walls during a discussion characterize some rooms. His designs of the workspace and in the robots he helped produce for Sony have been featured at the New York Museum of Modern Art. He is also an expert on high-performance computing and consulted with such innovative computer architecture companies as Thinking Machines.

HISTORY

Dr. Kitano has been working in systems biology since 1993 even while he was first starting at Sony. His current program was founded in 1998 by a grant from the Japanese government (Japanese Science and Technology Corporation, JST) as part of their ambitious ERATO (Exploratory Research for Advanced Technology) program. In 2000, the ERATO project officially became part of the newly created Systems Biology Institute, founded in 2000 to provide an entity that could accept funding from other sources so that the research seeded by ERATO could be stably funded. This Institute garnered follow-on funding from the ministry of agriculture and New Energy and Industrial Technology Development Organization (NEDO), a subbranch of METI (the Ministry of Economics, Trade and Industry). In 2003 the ERATO project finished, but with this extra funding and additional support from the Solution Oriented Research and Technology Program (SORST) in the Ministry of Education, the projects have expanded and continued.

The first round of ERATO funding targeted research on fundamental modeling and model standards technology such as the Systems Biology Workbench (SBW) and the Systems Biology Markup Language (SMBL), as well as new theory of the robustness of cellular networks and their application to yeast signaling. The current funding continues to support the work on standards and has expanded the research focus.

The project operates at approximately $2 million a year total cost, down from $3 million a year when the project started. However, that initial funding included research on humanoid robots that was spun out into a company and is not part of the current research program. This funding covers approximately six researchers including two full-time and one part-time researcher, one to two part-time graduate students, and one technician.

RESEARCH FOCUS

In addition to the broad areas above, Kitano outlined a number of other areas of research in the Institute including signal transduction in yeast and mammals (collaborating with the Alpha Project and the Alliance for Cellular Signaling), respiratory oscillations in yeast and calcium oscillations in mammalian cells (including collaborations with the Karolinska Institute), and advanced hardware platforms for simulations. Institute personnel discussed a few of these areas in detail.

Douglas Murray, a relatively new researcher at the Institute and previously of the Dynamics Group, Department of Biology, Beckman Research Institute of the City of Hope Medical Center, presented his research on the respiratory and reductive phase oscillation in the budding yeast, Saccharomyces cerevisiae. He has developed a highly controlled continuous culture system in yeast. The cycle he is studying was discovered in 1973 by Mochen and Pye and evolves after glucose repression followed by diauxic shift to become a respiratory oscillation with an approximately 40 minute period. In subsequent work, these oscillations have been shown to be linked to pathways as diverse as apoptosis, transcription and cell cycle and to be correlated with morphological changes in the cell. Other redox intermediates show oscillations in or out of phase with the aerobic respiratory response: nicotinamide adenine dinucleotide (NADH) oscillates in phase with O^2 usage and glutathione oscillates out of phase. Thus, a complex redox cycle is set up. Murray has related the oscillations shown here to the chronobiological processes of circadian rhythms. Like the circadian rhythm, this cycle seems to be temperature and partial pH compensated. Murray used Affymetrix-based gene expression analysis to follow cells synchronized via either an acetaldehyde pulse or H2S off-gassing. Interestingly, those cells synchronized using the acetaldehyde pulse showed both type 1 and 3 phase response curves whereas H2S perturbations results only in a type 1 curve. These are diagnostic of the structure of the oscillatory cycle as well as the role of these two substrates in the cycle. Using both imaging and microarrays, Murray demonstrated that the redox cycle not only gates the cell cycle but also regulates, in a coordinated manner, nearly all of gene expression (~9/10 of genes are expressed in the reductive phase and ~1/10 (mostly ribonucleic acid (RNA) synthetases) in the respiratory phase). The entire transcriptome, therefore, changes in roughly eight minutes. Using a Fourier focusing analysis, Murray discovered that appearance for the transcript for a process occurs about 20 minutes earlier than appearance of the phenotype. He then used Cytoscape to display his transcriptional results in coherent networks.

Dr. Hisao Moriya had just finished a post-doc in Mark Johnston's laboratory at Washington University when he joined the Institute in April 2004. His research concerns glucose sensing and conserved network structure in yeast as well as the yeast cell cycle. As he has just started and his research involves both experimental and computational components the results are preliminary. He used Kitano's Cell Designer, a graphical pathway design tool that explicitly represents molecular state and output pathways, to create a model of the hexose transporters induced by glucose exposure. In making the model he uncovered a number of feedback loops that seem to control when and which transporters are brought online. A particular configuration of positive and negative feedback loops found in this system also seems extant in control of both galactose and methionine systems, although currently no unifying theories exist as to the general purpose of this mechanism. Moriya is also exploring robustness in Tyson's cell cycle model.

Noriko Hiroi and Akira Funahashi are working on models to test alternative hypotheses about the way restriction enzymes move on DNA. This is an interesting system in which to study dimension-restricted motion and reaction, as there is a clear endpoint and geometry. Their target of study is EcoRV, a much studied *E. coli* restriction enzyme. They compare and contrast different mechanisms by which the enzyme discovers and cleaves its DNA target, including uncorrelated jumping, and more correlated hopping and sliding processes. They wanted a model (and data) of sufficient resolution to distinguish the hopping from the sliding process. They also wanted to contrast a stochastic process representation to a simpler stochastic model by Coppey et al. They built two separate models to represent different amounts of hopping and sliding of the restriction enzyme (based on kinetic parameters from A. Pingoud) and simulated the dynamics of cleavage that they could compare to existing data on cleavage rates measured by Stephen Halford. They were able to demonstrate that a correlated motion model was necessary to explain the data and to estimate the relative contributions of sliding and hopping. However, there is more validation to be done.

Akira Funahashi then described Cell Designer more fully (this is NOT an open-source tool yet) and a hardware-accelerated simulation system based on Field Programmable Gate Arrays (FPGAs). They have built a set of algorithmic design tools called RecSiP (Reconfigurable cellular Simulation Platform), which reprogram this software configurable chip for ordinary differential equation (ODE) or stochastic simulation. Though all recognized that much speed could be obtained just by algorithm design, the supposition is that whatever algorithm was programmed onto this chip would run much faster than a compiled language would on a more general purpose processor. Funahashi demonstrated that the system obtained an 18–19-fold increase for ODE-based simulations and over a 100-fold speedup for stochastic algorithms such as Dennis Bray's Stochsim.

Following these talks Kitano then summarized a number of other projects he is driving or collaborating in including a pheromone pathway modeling project with Mel Simon and Tau-Mu Yi in which they have tagged with green fluorescent protein (GFP) over 30 proteins in the pathway and obtained time-course and localization data under different knockout conditions. With this and other data they are trying to reverse engineer unknown parts of the pathway. Another, interesting project is the tracking of cells and nuclei during development of *C. elegans* from the fertilized egg through the 32-cell state. They have built automated image analysis systems and a type of 3D Nomarski imaging to localize the nuclei during cell division and track their motion during division. They can then correlate the motion of different nuclei and compare tracks of a given nuclei from multiple eggs to see how controlled and precise the process is.

When queried about educational programs, Kitano mentioned that Keio University has a relatively new department of biosciences and informatics with about 30–40 undergraduates. (http://www.st.keio.ac.jp/english/facu_bio/).

Site: **Kyoto University, Bioinformatics Center**
Institute for Chemical Research
Uji, Kyoto 611-0011, Japan
http://kanehisa.kuicr.kyoto-u.ac.jp/

Date: December 17, 2004

WTEC Attendees: F. Doyle (Report author), S. Demir, C. Stokes

Host: Minoru Kanehisa, Professor, Tel: +81-0774-38-3270, Email: kanehisa@kuicr.kyoto-u.ac.jp

OVERVIEW

The host (Prof. Minoru Kanehisa) spent a couple of hours with the group, primarily discussing the Kyoto Encyclopedia of Genes and Genomes (KEGG) database (http://www.genome.jp) project, as well as some curricular activities in bioinformatics. Prof. Kanehisa is the director of the Bioinformatics Center and is a former president of the Japanese Society for Bioinformatics. The Center is home to the well-known KEGG database. The Center includes six faculty and six instructors in the areas of bioknowledge systems, biological information networks, pathway engineering, proteome informatics, and genome informatics.

The KEGG project itself involves 30 full-time researchers. Prof. Kanehisa's group includes a total of 60 researchers at Kyoto, and an additional 13 at the University of Tokyo. A new lab will start up in Boston as well. KEGG has been primarily supported from government funding sources (mostly the Ministry of Education, Culture, Sports, Science and Technology (MEXT), also Japanese Society for Promotion of Science (JSPS), and Japan Science and Technology Corporation (JST)). The sizeable computational resources are provided by the Bioinformatics Center at Institute for Chemical Research, Universty of Kyoto. Operating costs for the KEGG project are $3 million a year (not including the computing).

As was explained, the KEGG database is quite different from those maintained by the National Center for Biotechnology Information (NCBI), notably in the ability to do reconstruction, and the retrieval of network features. As with some other database projects, they are beginning to address the chemical space (metabolites, glycans, lipids, etc.). The core elements of KEGG are the GENES database, the LIGAND database, and the PATHWAY element for network integration. The combination of both genomic space and chemical space allows both screening for target genes (e.g., disease genes), and screening for lead compounds and molecular probes, respectively.

Several research projects were described that implement sophisticated statistical and computational methods for network analysis and construction. The group relies heavily on graph theoretical tools (representations of binary relationships in the form of nodes and edges). Abstractions can be formulated that allow nested graphs (nodes of one graph are graphs). They are using such tools to do predictions (KEGG orthology) using manual and automated methods for curating orthologs. The computational algorithm has produced 30,000 orthologs over 200 organisms, and the manual approach has yielded 6,000 orthologs thus far. One of the more original studies in their group is the use of line graphs to integrate chemical and genomic spaces. This leads to a new method for chemical structure comparison that is akin to BLAST for genomic space.

They are addressing inter-operability issues, and have provided the KEGG markup language (KGML) version of their network models (http://www.genome.jp/kegg/xml/), which are used by both Systems Biology Markup Language (SBML) and genomic object net (GON).

Additional research in Prof. Kanehisa's group includes technology approaches to systems biology, such as scale-free network analysis, and kernel methods for network inference. Two projects were discussed in detail by two of his lab members (Kiyoki Aoki, and Jean-Marc Schwartz). These were, respectively, (i) glycan structure network models and structure search, and (ii) dynamic metabolic models using elementary flux modes. A recurring theme in his work, and an important one for systems biology, is the value of networks in generating insight into biological behavior. Although they are static (in this case), the utility of such information is very high, and can lead to effective screening mechanisms for disease-related genes, lead compounds and molecular probes.

There was an extended discussion of training in the area of bioinformatics, as Prof. Kanehisa was instrumental in formulating the curriculum that has been promoted by the Japanese Society for Bioinformatics. This curriculum can be viewed at http://www.bic.kyoto-u.ac.jp/egis/course.html. He estimates that there are approximately 10 programs in Japan that are bioinformatics-focused, with a couple outside of universities like the Computational Biology Research Center (CBRC). In addition, they have archived video of lectures that are stored online using the webCT course management system. They have strong relationshops with Humboldt University (Berlin) and Boston University with student internships and workshops.

There was a brief discussion of industry relations for the Center, and he noted that it is primarily through the training program (education as opposed to research). There are some interactions with individual companies, but an extensive contract relationship does not exist. They ask industry to pay for KEGG database downloads, but compliance is problematic.

Site: **Kyoto University, Cell/Biodynamics Simulation Project**
Kyoto 606-8501 Japan
http://www.biosim.med.kyoto-u.ac.jp/

Date: December 17, 2004

WTEC Attendees: C. Stokes (Report author), F. Doyle, S. Demir, G. Hane

Host: Dr. Akinori Noma, MD, PhD, Professor and Project Director, Department of Physiology and Biophysics, Tel: +81-75-753-4352,
Fax: +81-75-753-4349,
Email: noma@card.med.kyoto-u.ac.jp
Dr. Tetsuya Matsuda, MD, PhD, Professor, Department of Systems Science, Biomedical Engineering, Tel: +81-75-753-3372,
Fax: +81-75-753-3376,
Email: tetsu@i.kyoto-u.ac.jp

Professor Noma provided an overview of the Cell/Biodynamics Simulation Project, and two other project faculty, Dr. Tetsuya Matsuda and Dr. Nobuaki Sarai presented descriptions of their research and development efforts. About 15 additional group members (postdocs, students, and visiting researchers) also attended the discussion.

BACKGROUND

The Cell/Biodynamics Simulation Project (http://www.biosim.med.kyoto-u.ac.jp) is a national leading project for cooperation between industry and academia sponsored by the Ministry of Education, Culture, Sports, Science and Technology (MEXT). The project is focused on developing dynamic mathematical models of the cardiac myocyte and the heart as resources for cardiac research, industrial application and education. It has been in progress for about one-and-a-half years, and is funded for five years. This is a large collaborative project between Kyoto and Keio Universities as well as faculty at universities in Korea and Poland and, currently, nine Japanese companies. The research group in Kyoto is based in the research park area of Kyoto to allow all members of the group to have common office space instead of being separated in scattered offices within their home departments.

RESEARCH

The focus of the project is to develop a comprehensive cell model of the cardiac myocyte's functions. There is a long history of modeling of the membrane electrophysiology of cardiac action potentials and the sino-atrial (SA) node pacemaker, and, separately, cell models of intracellular ions, signaling, and metabolism. In this project, the aim is to integrate the two areas into a single model to have a better representation of the mechanisms regulating myocyte electrophysiology. The components of their model, the Kyoto model, were illustrated, and a demonstration of the model in the group's computer simulation platform was given. The Kyoto model to date focuses on membrane potential, ion channels, ionic balances, energy balance, osmolarity, cell volume balance, redox state balance, pH balance, and substrates and metabolites balance (related to adenosine triphosphate (ATP) utilization). There is not yet proteolysis or protein production. A major effort in integrating electrophysiology and intracellular biology has focused on the inclusion of ATP balance using a detailed model of mitochondrial function. Dr. Noma explained that model simulation results have been extensively compared to results from many different lab experiments in order to build confidence in the model's function.

Additional effort has been put into creating a simple whole heart model in order to include the effect of heart wall tension feedback onto myocyte function. The contraction of a single myocyte is used to represent the contractile properties of the whole "heart," and, given wall elasticity, the resulting heart pressure and volume feedback onto certain properties of the cell in a closed loop. Dr. Noma noted that this was much simpler than existing models elsewhere but it is a first step towards more comprehensive models of the heart.

Dr. Matsuda described his efforts to develop a more elaborate model of the heart using finite element modeling in conjunction with the electrophysiological cell models. He is investigating the shape change of the left ventricle when it beats, in particular the origin of the twisting that is seen. He has four-dimensional magnetic resonance imaging (MRI) data of the beating human heart that he is attempting to reproduce. Dr. Matsuda demonstrated a 5,000-element mesh representation of the left ventricle using the Kyoto cell model to represent the electrophysiology properties of each node and how his mechanical dynamics solver calculates dynamic shape change. The current model can reproduce the force-length relationship of isometric contraction quite well. He has investigated the different shapes that result through different orientations of myocytes in the ventricular wall. He noted that he has not yet found a myocyte orientation that accurately reproduces the ventricular twisting, and said that the investigation is ongoing.

The group has developed two modeling and simulation platforms, SimBio and DynaBioS®. Dr. Sarai described SimBio. It simulates the function of a single cardiac myocyte, and the Kyoto model is built and tested using this platform. The platform is written in Java, the models can be exported in XML, and it is available publicly on the web site (http://www.card.med.kyoto-u.ac.jp). The SimBio platform has numerous pre-programmed cellular elements, such as specific ion channels, that can be brought together easily to form a new model or modify an existing one. They have been organized to be easily accessible to biologists, not just modeling-savvy engineers.

DynaBioS is a modeling and simulation platform. With it they are constructing a beating heart simulator, combining a mechanical dynamics solver with an electrophysiology simulator as an initial example using DynaBioS. Dr. Matsuda explained that this software has a component-based architecture so it can utilize existing simulation software such as SimBio as sub-simulators. It is event-driven for asynchronous processing to facilitate large-scale parallel processing. This software is not yet available outside the group.

When asked why they developed their own modeling/simulation platforms rather than use existing platforms, they explained that they felt that existing software is focused on other applications and not well-suited to theirs, is computationally slow, and is difficult to use for the non-expert. Their goal is to create modeling/simulation software for electrophysiology applications that is easy to use for biologists. They are also developing a program which can convert a model described in Cell markup language (CellML) to Java and vice versa.

Research applications of the models are envisioned but, except for the study of the left ventricle, were described only briefly. It was explained that for many applications, the models must be developed further before they are complete enough for such work. Some work done to date includes investigating why knocking out a sodium/calcium exchanger in the heart in mice doesn't result in cell death; how mitochondrial function is coupled with work load; the cause-effect relationship between the membrane chloride ion fluxes, Na/K pump activity and the cell volume regulation; and the mechanism of action of certain cardiac drugs.

PERSONNEL

The project has more than 20 faculty members, 10 postdocs and a few graduate students. In addition, four people from companies also work in the group full-time as visiting researchers, helping to develop the platforms and models. The group is about half computer scientists/engineers and half biologists. See the section titled "Collaborations" for more information about the collaborating universities and companies.

BUDGET

The total five-year budget for the project consisting of three major centers including the Kyoto group is ¥8 billion, with about half of that provided in the first year to open the facilities needed for the lab. (http://www.lp-biosimulation.com/)

TRAINING

The training emphasis for project members is on learning by doing. The models themselves are used as teaching tools, where the students learn by building and using the models. Dr. Noma has discussed with an educational software provider the possibility of providing the models to others for educational purposes in some form, but nothing is in place to do so. He has used the models in his teaching at the university for over five years.

COLLABORATIONS

The Cell/Biodynamics Simulation is a large collaborative project between Kyoto and Keio Universities in Japan as well as researchers in Korea (at Ulsan, Kangwon and Cheju Universities) and Poland (at Jagiellonian University) and nine companies (currently). At Kyoto University the participating faculty are members of the Graduate School of Medicine, the Graduate School of Informatics, and the Graduate School of Engineering. At Keio University the participating researchers are faculty from the Institute for Advanced Biosciences. A major aim of the government in funding these centers is to stimulate the economy, so the project emphasizes collaboration with industry as well. Presently seven pharmaceutical companies are involved, with four each sending a full-time researcher to work on the project in Kyoto and three sending someone intermittently. The pharmaceutical companies are Nippon Shinyaku, Shionogi, Sumitomo Chemicals, Tanabe, Sankyo, Takeda, and Mitsubishi Well Pharma. These industry collaborations can be described as "development partnerships," where the benefit to the participating company is increased understanding of the methods, benefits and limitations of biological modeling as well as a general biological understanding derived from model development, but not ownership or dedicated use of the developed models for company research at this time. Dr. Noma said that only later would contracts be implemented with the companies to carry out specific research, specifying ownership and benefit of intellectual property that is developed.

Site: **RIKEN Yokohama Institute**
1-7-22 Suehiro-cho, Tsurumi-ki
Yokohama City
Kanagawa 23-0045, Japan
http://www.gsc.riken.jp

Date: December 16, 2004

WTEC Attendees: F. Katagiri (Report author: Sugiyama, Plant Science Center), M. Cassman (Report author: Sakaki, Protein Research Group), A. Arkin (Report author: Ueda, Bioinformatics group), F. Doyle, C. Stokes, S. Demir, G. Hane

Hosts: Tomoya Ogawa, Director, RIKEN Yokohama Institute
Tatsuo Sugiyama, Director, Plant Science Center, Email: sugiyama@postman.riken.go.jp
Shigeyuki Yokoyama, Project Director, Protein Research Group, GSC, Email: yokoyama@riken.jp
Akihiko Konagaya, Project Director, Bioinformatics Group, GSC, Tel: +81-45-503-9301, Fax: +81-45-503-9559, Email: konagaya@gsc.riken.jp
Toru Yao, Research Collaborative Advisor, Genomic Sciences Center, Tel: +81-45-503-9295, Fax: +81-45-503-9553, Email: yao@riken.jp
Kazuo Shinozaki, Project Director, Plant Functional Genomics Research Group, GSC; also Director Designee, PSC, Email: sinozaki@rtc.riken.jp
Taku Demura, Team Leader, PSC
Ken Matsuoka, Team Leader, PSC
Minami Matsui, Team Leader, Plant Functional Genomics Research Group, GSC
Motoaki Seki, Plant Functional Genomics Research Group, GSC
Yoshiyuki Sakaki, Director, Genome Sciences Center
Mariko Hatakeyama, Bioinformatics Group, GSC
Shuji Kotani, Bioinformatics Group, GSC
Jun Kawai, Team Leader, Genome Exploration Research Group, GSC

Hiroki Ueda, Head, Laboratory for Systems Biology, Center for Developmental Biology, RIKEN Kobe Institute

OVERVIEW

Ogawa presented an overview of RIKEN and the RIKEN Yokohama Institute. RIKEN has seven institutes in Japan and three international sites, RIKEN-Rutherford Appleton Laboratory Muon Facility in the U.K., RIKEN-Brookhaven National Laboratory Research Center in the U.S., and RIKEN-MIT Neuroscience Research Center in the U.S. The Yokohama Institute is the second largest institute among the RIKEN institutes and consists of the Genome Sciences Center (GSC), the Plant Science Center (PSC), the SNP Research Center (SRC), and the Research Center for Allergy and Immunology (RCAI). In FY2004, the budget for RIKEN was about $587 million, and the Yokohama Institute received approximately $156 million of this. Within the Yokohama Institute, nearly $80 million was allocated for GSC, about $16 million for PSC, about $21 million for SRC, and approximately $39 million for RCAI. Eight-hundred thirty-four staff members in the Institute are distributed among the centers approximately proportionally to their budgets. The Yokohama Institute has four research buildings, two nuclear magnetic resonance (NMR) complexes, and an office building. The total floor space of these buildings is about 60,000 m^2. The Yokohama Institute has a close relationship with the Yokohama City University, which is located adjacent to the Institute. The joint graduate program between the Institute and the University has approximately 150 students, and about a half of the teaching staff for the program are from the Institute.

RIKEN encourages its staff to commercialize discoveries and inventions in research. In the last six years, RIKEN Yokohama Institute has spun out 16 companies. They also have a corporate "Club," in which the Institute collects membership fees from member companies in exchange for early access to confidential information generated in the Institute. This club facilitates collaboration between industry and the Institute and offers licensing opportunities to the member companies.

GSC, headed by Yoshiyuki Sakaki, focuses on the genome and performs comprehensive research on the structure and function of animal and plant genes, genomes, and proteins, from the molecular to the organismal level. GSC consists of six research groups, including the Genome Exploration Research Group, led by Yoshihide Hayashizaki, the Protein Research Group, led by Shigeyuki Yokoyama, the Plant Functional Genomics Research Group, led by Kazuo Shinozaki, and the Bioinformatics Group, led by Akihiko Konagaya, and the Genome Core Technology Facilities. A large international collaboration, RIKEN Structural Genomics/Proteomics Initiative (RSGI), is led by Yokoyama, and RSGI's goal is to determine the structures of 3,000 proteins in five years ("Protein 3000 Project"). NMR complexes at the Yokohama Institute and an X-ray source from a synchrotron at the RIKEN Harima Institute are being used for NMR-based and X-ray-crystallography-based structure determination, respectively. The structures of 2,500 of the 3,000 proteins will be determined at RIKEN.

Sugiyama, Director of PSC, gave an overview of PSC. The objectives of research at PSC are: (1) deepening the molecular understanding of the higher order function of plants using model plants by elucidating gene function and bio-molecular behavior ("Learn from plants") and (2) social usage of the basic fruits along with the Millennium Rice Genome Project for better living and eating ("Utilize plants"). PSC consists of six research groups, including the Morphogenesis Research Group, to which Demura and Matsuoka belong.

Demura presented his work on gene discovery for xylem cell differentiation. They use an *in vitro* culture system using *Zinnia elegans*, in which mesophyll cells differentiate into tracheary elements (TE) in three days. They also developed a similar *in vitro* culture system using *Arabidopsis* for a cell differentiation study. The messenger ribonucleic acid (mRNA) level changes during the process in both *Zinnia* and *Arabidopsis* were monitored using microarrays. Candidate genes for regulation of this differentiation process were identified based on the microarray results and systematically screened by over- and under-expression of the genes in *Arabidopsis*. An overexpression of the *Arabidopsis* gene X ectopically converted epidermal cells into TE. When gene X was underexpressed, protoxylem (but not metaxylem) was lost.

Matsuoka presented expression profiling analysis of the tobacco cell line BY-2. The cell line is widely used due to its amenability. Candidate genes identified by expression profiling for various biological phenomena are tested by overexpression of the genes in the cell line. In a case study, a methyl jasmonate inducible Myb gene was overexpressed. The overexpressing line accumulated phenolics, which is similar to methyl jasmonate-treated cells. The Myb gene is hypothesized to function in induction of phenolics synthesis genes in response to methyl jasmonate. One of the challenges for the research is how to translate the results into crop improvement.

The results of these projects are available at http://mrg.psc.riken.jp/.

Shinozaki, who is becoming director of PSC, presented *Arabidopsis* functional genomics research in his group at GSC. To take advantage of the genome sequence information in *Arabidopsis*, Shinozaki's group generated two types of resources: a collection of tagged *Arabidopsis* mutant lines and a collection of full-length complementary deoxyribonucleic acid (cDNA) clones (RIKEN *Arabidopsis* full-length (RAFL) cDNA). Approximately 70,000 activation-tagged lines and about 18,000 Ac/Ds transposon-tagged lines were generated. About 18,000 Ds insertion sites of the tagged lines were determined. Phenotypes of about 5,000 Ac/Ds lines were analyzed. The phenotyping methods include the work of Matsui with laser range finder, which captures 2D-depth images of a plant, and X-ray micro computed tomography, which captures 3D images. These shape phenotypes are statistically analyzed by computer to extract quantitative traits using shape models. The RAFL cDNA project was detailed in Seki's presentation. Approximately 18,000 RAFL cDNA clones, which correspond to 60–70% of the total genes in *Arabidopsis*, were isolated. The clones were used to make a cDNA microarray. They are being used to produce proteins for protein structure determination, in collaboration with Yokoyama's group. The clones will be used to generate over- and under-expression transgenic lines. The above information is organized in the RIKEN *Arabidopsis* Genome Encyclopedia (RARGE) database and is available at http://rarge.riken.go.jp/.

In the next phase, they are aiming to integrate genome to phenome and metabolome. The tagged lines will be analyzed to collect transcriptomic, phenomic, and metabolomic data. Bioinformatics will be crucial for understanding the resulting complex data. In addition, they will generate full-length cDNA collections from crops and trees. These clones will be used to functionally analyze the genes in *Arabidopsis*.

Genomic Sciences Center Director Dr. Sakaki presented the programs of the Center. The GSC was established as the leading center of genome sciences in Japan in 1998, using the successes of the Human Genome Project as an incentive. The ultimate goal, defined by Dr.Wada in 2003, is to integrate genome and phenome through a comprehensive informatics program incorporating the intermediate steps of transcriptome, proteome, and metabolome. The GSC includes seven separate research groups, of which the Protein Research Group is the largest. This group is responsible for the Japanese Structural Genomics effort, and, together with the Bioinformatics Research Group, is described below. Other research groups are responsible for the sequencing of chromosomes 11, 18 and 21; functional annotation of mouse (FANTOM) full-length cDNA clones, which has generated the world's largest clone bank; and a mouse functional genomics research group which has carried out large-scale N-ethyl-N-nitrosourea (ENU) mutagenesis, creating 20,000 mutant alleles with point mutations. Twenty-four of these mutants have phenotypes corresponding to human diseases. Additionally, a genome network project was begun in October, 2004, to identify transcriptional regulatory networks in the human genome. This is a national project, where the biological projects and technology development is performed by 12 groups of independent investigators while RIKEN provides the core data and resource production capability. RIKEN is conducting a functional analysis of promoters and transcription factors using both two-hybrid approaches and a protein chip (X-chip) to create a protein-protein interaction network.

The GSC has a number of specific projects underway using either network approaches or technology developed at RIKEN to examine specific biological questions. One involves screening drug effects on circadian rhythms, using proposed networks identified as being involved. They have also constructed a high-throughput screening capability that can address genotyping, protein structure analysis, target selection, and drug design. Ultimately, the intent is to go from protein-protein interaction networks to gene regulatory networks to cell determination, with the ability to predict inhibitor effects by understanding the mechanism of protein-protein interaction and its effect on the network(s). This is expected to greatly aid drug discovery. Finally, a group at GSC has developed a special-purpose chip with petaflops performance that can be used in molecular dynamics simulations particularly for *in silico* ligand docking on proteins of medical and biological interest. This will be effective for identification of lead compounds and their optimization.

This very ambitious program is designed to reach across the gap that exists between the genome and cellular (and organismal) phenotypes, using network assembly, computational approaches, and simulation to generate a fundamental understanding of biological systems and consequently to provide new approaches to drug discovery.

THE PROTEIN RESEARCH GROUP

This group is directed by Dr. Shigeyuki Yokoyama and has as its primary role the development of structural genomics. The intent of structural genomics is to systematically analyze structure-function relationships of all the proteins in the genome(s). By identifying families of proteins through genomics it may be possible to create a systematic approach to modeling proteins of unknown 3D structures and to infer structure-function relationships.

The "Protein 3,000 Project" started in 2001 and is part of an international effort in structural genomics. It has a budget of $45 million/year at RIKEN, with a total of $90 million when including ancillary activities at other sites. However, the primary activity is at the NMR resource at RIKEN Yokohama Institute and the SPring-8 synchrotron resource at the RIKEN Harima Institute. There are two beam lines for the use of this project at Spring-8. Crystallization and sample handling have been fully automated, as well as data collection. The NMR resource focuses on proteins of less than 20–30,000 molecular weight. This resource is unique in the world. It contains two 900 Mhz, eleven 800 Mhz, six 700 Mhz, and seventeen 600 Mhz machines. Efforts to fully automate structure determination exist at the NMR site as well. The NMR is also used for screening folded proteins.

One aspect of the project is the ability to get the structures of protein-protein complexes. Identification of the interactions, particularly in transient cellular complexes, will be valuable in providing a mechanistic and structural basis for the formation of intracellular networks.

A major issue in such large-scale efforts is a rapid and easy method for the acquisition of protein. The RIKEN group has used a unique cell-free protein synthesis system that can produce material without the excess peptides found in most recombinant proteins. This system with detergents can also produce integral membrane proteins. A green fluorescent protein (GFP) fusion is used to monitor folding in 96 well systems: if the protein folds properly, the fused GFP also folds properly and the fluorescence from GFP can be detected. The cell-free system permits the preparation of labeled protein, in the order of milligrams, necessary for NMR, as well as the semen labeling necessary for X-ray determination.

Since its initiation the project has completed more than 500 protein structures, roughly 50:50 X-ray:NMR. The focus is on human cDNA, with targets based on disease relationships, drug targets, signal transduction proteins, etc. They are also collaborating with industry to do virtual screening of drug-protein interactions. Four companies are now involved in proprietary research. In addition, another project is underway using *thermophilus*. This organism has only about 2,000 genes, and it is planned to get a minimum set in the next two years.

YOKAHAMA RIKEN BIOINFORMATICS GROUP

Bioinformatics Director Akihiki Konogaya gave the overview of systems biology for this team. The mission for the group is information integration from the genome to the phenome (all the phenotypes of a cell or organism), mathematical modeling and simulation, high-performance computing for molecular dynamics and other biosimulation, and support for many of the large genome projects (such as full-length long cDNA collection of Japan (FLJ) and PHANTOM) in Japan. The bioinformatics group has five teams: The immunoinformatics team, the cellular knowledge modeling team, the population and quantitative genomics team, the genomic knowledge base research team, and the high-performance biocomputing research team. A computer and information facilities group manages grid and web services.

Konogaya defines a system as a well-organized superstructure of components, featuring new functions that are not shown in the components alone. The key features of such systems are mechanisms of regulation that process signals and maintain the coherence of the system and the dynamics that define its spatiotemporal behavior. The center is directly approaching key issues in modeling such systems. These include the granularity of the system description; whether the predictions should be qualitative or quantitative; whether the description should be stochastic or deterministic, spatial or homogeneous, etc. They admit that no "royal road" to biological modeling exists and that each system and question about the system admits a different level and approach to modeling. They have a number of examples of successes of different approaches. For example, they have used logical inference on causal "triad" models (where all interactions are represented by object, output, effector relations) and a knowledgebase of drug interaction to infer the mechanisms of that interaction. They have projects in differential equation modeling of the cell cycle in yeast, and signal transduction in the ErbB pathway in mammalian cells (more on these below). They have developed novel Monte Carlo particle dynamics simulations for reaction-diffusions systems and a new petaflop processor specialized for molecular dynamics and related simulation, which can simulate systems of up to 380,000 particles (MDGrape3). They have also developed computational infrastructure to support high-end dissemination of data and for distributed and Grid computing. This framework is called the Open Bioinformatics Grid (OBIGRID; http://www.obigrid.org). The computational system has about 363 nodes with 619 central processing units (CPUs) provided by 27 sites. With this system they were able to demonstrate a simultaneous fit of 1,000 parameters of a signal transduction model at once on a system that could scale to 1,000 processors. A few in-depth examples of these efforts were then discussed.

Systems Biology of ErbB Receptor

Mariko Hatakeyama, a signal transduction biologist, presented her studies on the modeling of pathways surrounding ErbB-mediated signal transduction. She then used the model to explain data on how mutation and overexpression of this class of receptor leads to cancer and other defects. Different isoforms of ErbB (an epidermal growth factor (EGF)-type receptor) show different phenotypes under different conditions. The model they built contains events from the initial homo- and heterodimerization of receptors upon ligand binding all the way to expression of genes in response. Their models were run on their own network simulator, called Yet Another Gene Network Simulator (YAGNS). Since there was a great deal of prior work on ErbB1 they focused their efforts at elucidating the specifics of ErbB4. This project is an excellent example of an experiment/modeling cycle.

In their initial study, they compared the data derived from quantitative western blots. These were used to follow the activity of 10 proteins in time (six time points) after ligand exposure. Fifty parameters of the model were fit to the data using the genetic algorithm- (GA) based estimator provided by YAGNS. They were particularly interested in the effect of a new interaction they had experimentally discovered (the action of a PP2A phosphatase). They were able to fit all the time courses (which may not be surprising given the large number of parameters fit and the paucity of data) and then predicted the effect of over- and under-expression of PP2A phosphatase. They found that repression or low activity leads to loss of Raf/AKT cross-talk. This prediction was tested experimentally using a PP2A inhibitor. The results were different in cells expressing ErbB1 or both ErbB1 and ErbB4. To explain this difference they hypothesized a new interaction in the ErbB1/4 cells involving a new PLCβ, EAP1, in the B-raf pathway activating mitogen-activated protein-ERK kinase (MEK) specifically in ErbB1/4 cells. This was validated experimentally.

Going to finer granularity they then used their molecular dynamics hardware engines, utilizing the petaflop processor, to explore how differential phosphorylation of various tyrosines on the ErbB receptor could control different signal transduction protein bindings and thus differential pathway activation. In particular they looked at whether molecular dynamics simulations could explain Grb2 domain specificity for different phosphoproteins. In an article published in the Journal of Biological Chemistry (2004), they show an excellent linear correlation of the predicted specificity of the phosphoinositide 3-kinase (PI3K) p85 SH2 domain to experimental results.

Based on this study they continued the study of the PI3K p85 SH2. They used a homology model for each of the ErbBs to show that there was a novel binding mode for the kinase to the receptor. They showed that the different phosphorylation states of the receptor could affect this binding mode. Because of the good experimental correlations, the predictions of the MD models can be added as parameters to the ODE models discussed previously.

Additionally, they are trying new approaches to infer networks from gene expression networks. In particular, using a clustering algorithm of expression data over a number of conditions and time, they break the system into a set of "modules." They then express the dynamics of the "modules" using Savageau's S-system representation starting with a fully connected network. They then fit the representation to the time-series data. This method currently works to estimate the dynamics of thirty variables. They tested their algorithms using data in the expression of the ultraviolet response (UVR) system in *Thermus Thermophilus*. The model was derived from data on 612 putative open reading frames (ORFs) in the system from which 24 clusters were derived with one cluster set aside for the uvrA gene. Their data included 14 time points following UV exposure of either wild type or an uvrA distruptant. The stochastic algorithm was run 10 times and obtained the same answer nine out of ten times. From this analysis they found a number of "key" clusters of unannotated genes that probably play a role in energy metabolism. They are now helping the receptor tyrosine kinase (RTK) consortium (http://receptorkinase.gsc.riken.jp/index.php) to use these methods in the study of mammalian systems.

Mammalian Cell Cycle Modeling

Shuji Kotani presented work on using Miyano's hybrid Petri net simulator and model building tools to explore cell cycles in *Saccharomyces cerevisiae*. The model included all four stages of the cell cycle, the nine cyclin dependent kinases (CDKs) and 14 cyclins that regulate them. All the phosphorylation, dephosphorylation, inhibition and degradation reactions were included. Using this model they predicted that cells overexpressing securin would arrest in G1. They also predicted that if Wee1 were over- or under-expressed by 30 times there would be an arrest in G1 but if the expression change were just 20 times then the cell time would decrease in the latter and increase in the former expression regime. Further, the model predicts that as Emi1 increases, the number of cell divisions before arrest decreases. In all three cases, experiments are in quantitative or qualitative agreement with the model. How the model was broken into discrete and continuous parts and where all the parameters came from was not clear.

Circadian Clock

Hiroki Ueda, MD, PhD, is a young researcher who recently left the Yokahama RIKEN for one in Kobe. He has worked with Kitano and others and has been recognized since his early graduate years as a leading researcher in systems biology. In fact, as a student he was given his own laboratory at RIKEN. Throughout his career he has been interested in quantitative and systems biology and, in particular, in the molecular basis of circadian rhythms, especially in mammals. His lab is now composed of 12 people: 1/3 dry lab, 2/3 wet lab. He believes this ratio is optimal because it takes more hands to do experiments than theory. There are three researchers (postdocs), four technical staff, four students and one assistant.

Ueda defines systems biology by four major efforts: systems identification (network inference, etc.), systems analysis (e.g. modeling), system control (i.e. using models and other methods to determine how to move a system from state A to B), and system design. He has studied the circadian clock as an excellent test bed for each of these efforts.

To start this effort he has focused his work on experimental system identification. He starts with genome- wide gene expression data, which he then clusters. From these clusters he uses sequence analysis (and known clock transcription factor cis-regulatory sequences) to find putative regulatory boxes. For those genes with strong predictions for the presence of these boxes and which seem to be clock controlled, he then builds promoter fusions to a version of luciferase (destabilized to a 20–30 minute half-life with a PEST tag) to follow their dynamics in RAT-1 fibroblasts.

Using this approach he has identified 101 clock-controlled genes in the suprachiasmatic nucleus (SCN) and 395 genes in the liver. Using these genes he then found promoter elements by clustering profiles followed by alignment of the upstream sequences and through comparative genomics using alignments with orthologs in human, mouse and rat. Thus, each gene becomes associated with a set of cis-regulatory sequences (in this case three of them: the E/E'box, the D box and the rev-response element (RRE) box). The challenge then is to correlate the pattern of appearance of these sequences upstream of a gene with the pattern of expression of that gene during the cycle. He was able to classify that presence of E/E' led to "day time expression," presence of D and RRE led to "night patterns," RRE and E/E' to evening patterns. Using the luciferase methods above he was able to verify the *in vivo* data using clock genes with wt and mutant cis-regulatory regions and stimulation of dexamethasone.

With all this work he was able to show that there were some evening genes regulated by D-box promoters, found temporal separation of the activators and repressors for the promoters, and was able to distinguish among three hypotheses about how this separation controlled the clock for each of the interacting systems (E, RRE and D). He elucidated how all the cross-regulation of the different sets of genes occurred and made a conceptual model sufficient for predicting that E was the critical regulator of the cycle and thus would be the best place of attack to change the cycle. He tested this prediction by expressing an E/E' box repressor and proved that the oscillation is lost in a dose-dependent manner whereas overexpressing an RRE or D repressor did little to stop the cycle.

Given these successful efforts he is now attacking the wider signal transduction network by using better microarrays (three times more genes have now been found); using small interfering ribonucleic acid (siRNA) and the copy deoxyribonucleic acid (cDNA) clone library for overexpression to find direct and indirect regulators of the three phases of gene expression; using knockout mice for similar reasons and for behavioral experiments, and using tagged clones to detect interaction by mass spectroscopy (he already has preliminary results on this). To aid in this research he is also developing new technology for measurement including a high-throughput phenotype analysis system that can simultaneously monitor twelve 24-well plates for bioluminescence and a new behavior analysis room where they can follow 300 mice simultaneously.

Encyclopedia of Genome Networks

Jun Kawai presented the encyclopedia of full-length cDNA transcripts that is now being extended to networks. Dr. Kawai is the team leader for the genome informatics exploration team and is representing the overall group leader for the genome exploration research group, Yoshihide Hayashizaki. For the last six to seven years this team has been building the RIKEN Mouse Genome Encyclopedia. This project is part of a consortium collecting and characterizing the full-length cDNA clones covering the full mouse transcriptome. They collect the physical clones, gene expression profiles and protein-interaction data generated by members of the FANTOM (Functional Annotation of Mouse cDNA) project and at RIKEN. They also aid in directed annotation of these data. To obtain the sequences of these clones they have a custom-built 384 capillary sequencers (the RIKEN integrated sequence analysis (RISA) sequencer) capable of 2,000 samples/day. They have 20 such machines at their site. As part of the FANTOM group they hosted the FANTOM jamboree (some of which the team heard about at JBIRC). They have documented more than 51,000 genes, of which a large number are alternative splice forms, and nearly half of which are non-coding RNA.

To ship these DNA clones (67,770 of them) in tubes takes 100kg of dry ice and numerous boxes so the center developed a new technology called DNA notes in which the clone, in a plasmid, is dried on paper. The entire collection is now in a single book worth about $1,000. Plasmids may be recovered by cutting out the paper and amplifying the plasmid directly from it.

In 2004 the genome network project started. They will expand their protein-protein, protein-DNA and protein-RNA interaction determination and expression correlation efforts and integrate this data with their genome databases. This involves large cross-institutional, even international, effort. This involves a highly complex organizational structure. But this Center in particular, apart from aiding in the informatics, is also developing new technologies for identifying transcriptional start sites (Cap analysis of Gene Expression) and for identification of promoter elements and transcription factor interaction.

The service to systems biology of genome production efforts like this, and the clone/strain libraries for *E. coli* at Tomita's operation, and Kodama's antibody production effort, is to provide an open-source, easily usable resource for tagging, overexpressing and knocking down biomolecules. All of these are essential for testing models of these systems.

Site: **Tokyo Medical and Dental University**
Center for Information Medicine
Yushima 1-5-45, Bunkyo, Tokyo 113-8510
http://www.tmd.ac.jp/mri/mri-end/index_j.html

Date: December 15, 2004

Attendees: M. Cassman (Report author), A. Arkin, F. Doyle, F. Katagiri, C. Stokes, S. Demir, R. Horning

Hosts: Hiroshi Tanaka, Director General and Professor, Tel: +81-3-5803-5839, Fax: +81-3-5803-0247, Email: tanaka@cim.tmd.ac.jp,
Hiroshi Mizushima, Associate Professor, Tel: +81-3-5550-2020, Fax: +81-3-5550-2027, Email: micom@tmd.ac.jp, Url: http://www.mizushima.info/hiroshi/,
Yoshihito Niimura, Associate Professor, Tel: +81-3-5803-5840, Fax: +81-3-5803-0247, Email: niimura@bioinfo.tmd.ac.jp,
Masatoshi Hagiwara, Professor and Director, Tel/Fax: +81-3-5803-5836, Email: m.hagiwara.end@mri.tmd.ac.jp,
Yasuhiro Suzuki, Assistant Professor, Tel: +81-03-5803-5840, Fax: +81-03-5803-0247, Email: suzuki.com@mri.tmd.ac.jp,
Soichi Ogishima, Doctorate Course Student, Department of Bioinformatics, Tel: +81-3-5803-5840, Fax: +81-3-5803-0247, Email: ogishima@bioinfo.tmd.ao.jp

OVERVIEW

Dr. Tanaka introduced the program. He is an engineer by training, but with a background in bioinformatics as well. He is the director of the Center for Information Medicine, professor, department of bioinformatics, and professor, department of systems biology. The Center has eight faculty members, two postdocs, and 25 graduate students. The primary research interests of the Center are systems evolutionary biology and genome medical informatics.

Systems evolutionary biology is focused on the understanding of gene and signaling networks based on evolutionary approaches. The assumption is that evolution consists of an elaboration in the processes by which information is transmitted through biological networks. Using this approach, he and his group have been investigating the Hox gene cluster responsible for defining the body plan. He is also interested in the development of protein-protein interaction networks and their evolution. He proposes that protein interaction networks have a hierarchical structure. By measuring cluster coefficients, he stratifies the systems into three layers; a large number of nodes with few interactions, a small number of nodes with many interactions, and an intermediate layer, which is assumed to be the most important in communication. Ogishima, a graduate student, presented some studies that measured degrees of protein-protein interaction in *E. coli*, yeast and *thaliana*, suggesting that these interactions evolve through modular formation and that the proliferation of an "intermediate layer" of Hox genes coincided with an evolutionary explosion of body plans. Suzuki reported on studies of the interaction of complexes in yeast, and on the p53 regulatory system. Finally, Niimura presented a discussion of the evolution of olfactory receptors in humans and mice. This is the largest known multigene family, and comparisons of human and mice suggested both a gain of genes in mice and a loss of genes in humans.

The meeting concluded with an open discussion between panel members, faculty, and students.

Site: **University of Tokyo, Department of Computational Biology**
Graduate School of Frontier Sciences
5-1-5 Kashinoha, Kashiwa City
Chiba, 277-8562
Tokyo, Japan

Date: December 15, 2004

WTEC Attendees: F. Doyle (Report author), A. Arkin, S. Demir, C. Stokes

Hosts: Shinichi Morishita, Professor, Tel: +81-47-136-3984, Fax: +81-47-136-3977, Email: moris@k.u-tokyo.ac.jp
Takashi Ito, Professor

OVERVIEW

The host (Prof. Shinichi Morishita) organized a session to review his work in the area of computational biology. In additional to himself, there were presentations by Prof. Takashi Ito (the chair of the department), and two research associates (Mr. Masahiro Kasahara and Mr. Tomoyuki Yamada). Their department is only two years old, and they have six faculty members (five in computer science and bioinformatics, and one wet biologist (Prof. Ito)). This is the first official Department of Computational Biology in Japan, although others are changing their names. They believe that a demand exists for PhD students in this area, and this is validated by companies that want to send their employees to the program to obtain a PhD. They have approximately 30 graduate students in the program (MS and PhD). They teach the graduate-level courses in their facility, and the undergraduate courses are taught on another campus of the University of Tokyo. Their undergraduate course attracts students majoring in science disciplines (biochemistry, physics, computer science, mathematics), and focuses on bioinformatics, although the students do actual sequence collection. Other topics include algorithms and design, genome assembly, protein structural analysis, and expression analysis. They are looking to extend the education activities with videoconferencing. There was some general discussion of the environment for start-up companies at their campus, and in Japan. They indicated that the Ministry of Education is encouraging start-ups from universities, and is aiming for 1,000 new companies over the next 10 years in Japan. The University role in this area is two-fold: they assist with patent applications and they connect the faculty with VC companies.

Prof. Ito outlined his own research interests in (i) functional genomics and systems biology of budding yeast, and (ii) mammal epigenomics. They are using multiple sources of quantitative data (proteomics, transcriptomics), and are interested in adding dynamic data and exploiting new developments in mass spectrometry (for proteomics). Additional assays include metabolites (using a fluorescence resonance energy transfer (FRET)-sensor), TF-binding (chromatin immunoprecipitation technique (ChIP) and generalized adaptor-tagged competitive protein-coupled receptors (GATC-PCR)), phosphorylation (mass-spec), and protein complex (mass-spec). The particular system of interest is the GCN pathway in budding yeast. They have a preliminary model (in Genomic Object Net (GON) format). A noteworthy result is the observation of oscillations in the model that have not yet been described in the literature. The model is used to drive new quantitative measurements. They collaborate internationally with the Yeast Systems Biology Network (http://www.ysbn.org).

Prof. Morishita describes his research program, broadly in the area of computational approaches to Omics. Specific project area that were detailed (by both Prof. Morishita and his students) included: (i) whole genome shotgun assembler, (ii) effective and target specific small interfering ribonucleic acid (siRNA) design, (iii) highly specific short oligomer design for measuring absolute abundance of mRNA, (iv) high-throughput start site collection, and (v) image processing for assembling morphological changes by mutagenesis. Focusing on the last project, they have developed effective image processing tools for analysis of two model systems—budding yeast and Drosophila wing development. In that work, they are looking to link specific gene mutations to the corresponding morphologic changes. On the yeast problem (a collaboration with Prof. Ohya at the University of Tokyo), they have collected data from 2 million cells, representing 5,000 disruptants. They focus on cell wall location, nucleus location, and actin distribution, using 500 distinct parameters to capture these effects in the images. They find that half the disruptants have significant morphological deviations. On the Drosophila wing morphology work (collaboration with Prof. Aigaki at Tokyo Metropolitan University), they have identified 135 parameters from 20,000 images of 3,000 mutants, and they find that 2/3 of the mutants show significant morphological alterations. Their overall goal is to use the Drosophila work as a model for human disease systems.

The group relied heavily on computers for their research, and run their own multi-node cluster machines with typical specifications of 64 CPUs and 128GB of the main memory.

There was some discussion about the role of industrial collaborations, and research funding in general. Prof. Morishita runs a group with four research associates and eight graduate students. They manage their own administration of their extensive computing facilities. Approximately 90% of his funding is from the government (Ministry of Education, Culture, Sports, Science and Technology (MEXT)—Grant-in-Aid Program for Scientific Research on Priority Areas; Japan Science and Technology Corporation), and 10% from industry (mostly gift funding from companies like Sony, Hitachi, and venture capital firms (VCs)). In return the companies get a first right to refusal on intellectual property (IP). He manages three large-scale computing clusters, and these represent approximately 2/3 of his annual research operating expenses. Prof. Morishita has helped to establish a start-up company to design siRNA sequences (RNAi Corporation).

Site: **University of Tokyo, Institute of Medical Science**
Human Genome Center
4-6-1 Shorokanedai
Minato-ku
Tokyo 108-8639, Japan
http://www.hgc.ims.u-tokyo.ac.jp

Date: December 13, 2004

WTEC Attendees: F. Doyle (Report author), A. Arkin, M. Cassman, S. Demir, R. Horning, F. Katagiri, C. Stokes

Host: Satoru Miyano, Professor, Tel: +81-3-5449-5615, Fax: +81-3-5449-5442, Email: miyano@ims.u-tokyo.ac.jp

SUMMARY

The host (Prof. Satoru Miyano) organized a session to review his work in the area of computational biology. In additional to himself, there were four others from his lab in attendance, including graduate students and postdocs. Dr. Miyano's background is in mathematics and computing theory, and he is currently the president of the Japanese Society for Bioinformatics, and he is the editor-in-chief of the newly established IEEE Transactions on Computational Biology and Bioinformatics. He outlined his research interests in the following areas: (1) Estimating gene networks from genome-wide biological data, (2) software tools for bioinformatics, and (3) pathway database projects.

In the area of network inference, he has been active with yeast networks since 1996. He has done 500 knockouts of 6,000 genes in yeast. His algorithmic approach to network inference involves Bayesian networks and non-parametric regression. To determine the optimal structure, they employ an information theoretic cost criterion with a penalty for model complexity. Although this is a nondeterministic polynomial time (NP)-hard problem, they have exploited computer science heuristics and high-performance computing to develop efficient solutions. This gene network inference method can yield the optimal gene network for 20 genes (on a Sun Fire 15,000 100 CPU machine) in one day. A test case ("success story") was described for the oral antifungal Griseofulvin. They measured gene expression as a function of exposure, and used a Boolean network approach with the drug as a virtual gene. This led to a prediction of the genes that were directly affected by the drug. They ultimately showed that CIK1, the primary affected target gene, is knocked down and that a knocked down strain shows the same phenotype as the drug (mitotic spindle formation). This work was done in collaboration with a small pharmaceutical company, and they were able to find pathways that were affected by the company's drug, suggesting that drugs that affected related components in the pathway would also be effective.

There was an extensive discussion of the group's software tools, including multiple demonstrations. In later visits, the WTEC team learned that many groups in Japan are using these tools. Effectively it is a Petri net approach, with a hybrid extension that allows combinations of discrete and continuous behavior (http://www.genomicobject.net). The tool is Java based, and the package is called Cell Illustrator (current version 1.5). There are several case studies included with the release, and a trial version (with timeout) is freely available. An adjunct package, Cell Animator, allows cartoon animations of the dynamic simulations. They have developed a markup language (Cell System Modeling Language—CSML) that can import Systems Biology Markup Language (SBML) level 2 code, as well as CellML code. In their latest work, they are developing converters to take major database networks (KEGG, Biocyc, WIT, BRENDA, etc.) and generate CI models. Ongoing efforts are looking at conversion of signaling pathways (e.g., Transpath) into CI models.

There was some discussion about research funding support for his group, and in Japan in general. He reported that he runs a lab of about 20 people on approximately $1 million/year in federal support. (He pointed out that many of his graduate students are "volunteers" and are not paid very much.) About $300,000/year comes from the Human Genome Center, and the balance is from smaller federal grants. His computing resources are provided by the Human Genome Center, which receives about $10 million/year for its operation. He receives some additional support from his own start-up company (Gene Networks International—http://www.genenetworks.com), although he noted that there are conflict of interest issues concerning federal funding and support from his company (intellectual property issues). Through the company, he conducts wet lab collaborations with colleagues at Kyushu University and Cambridge (U.K.).

There was also some discussion concerning formal education programs in Japan. In their own program (bioinformatics), they do not have a graduate program, although the Japanese Society for Bioinformatics has developed a curriculum. They collaborate with the Kyoto Bioinformatics Center in that regard. An international collaboration involves Boston University, the Kyoto Bioinformatics Center and Humboldt University in Berlin that includes meetings and student exchange.

Site: **University of Tokyo, Laboratory of Systems Biology and Medicine (LSBM)**
Research Center for Advanced Science and Technology
4-6-1 #34 Komaba, Meguro-Ku
Tokyo 153-8904, Japan
http://www.lsbm.org/site_e/index.html

Date: December 13, 2004

WTEC Attendees: C. Stokes (Report author), M. Cassman, A. Arkin, F. Katagiri, F. Doyle, S. Demir, R. Horning

Host: Dr. Tatsuhiko Kodama, Director, Department of Systems Biology and Medicine,
Tel: +81-3-5452-5230, Fax: +81-3-5452-5232,
Email: kodama@lsbm.org
Sigeo Ihara, Professor, Tel: +81-3-5452-6539,
Fax: +81-3-5452-6578, Email: ihara@lsbm.org
Dr. Hiroyuki Aburatani, Professor, Tel: +81-3-5452-5352, Fax: +81-3-5452-5355,
Email: aburatani@lsbm.org
Dr. Takao Hamakubo, Professor, Department of Molecular Biology and Medicine,
Tel/Fax: +81-3-5452-5231,
Email: hamakubo@lsbm.org

An overview of the laboratory and some of the technology development was given by Dr. Kodama. His colleagues who also attended, Dr. Sigeo Ihara, Dr. Hiroyuki Aburatani, and Dr. Takao Hamakubo, described research projects.

BACKGROUND

The Laboratory for Systems Biology and Medicine (LSBM; http://www.lsbm.org) is part of the Research Center for Advanced Science and Technology (RCAST) and is focused on both basic and applied systems biology research. LSBM has a basic research thrust that Dr. Kodama termed "reverse systems biology," and an applied research thrust that he called "genome-based antibody diagnostics and therapeutics." The research is focused on lab- and data-driven hypothesis testing, with an emphasis on genomics- and proteomics-based data-generation and analysis. There is much statistics-based data analysis but little or no mechanistic or dynamic modeling. There is a significant emphasis on the development of antibodies for therapeutics and antibodies, and this, as well as the basic research, is supported by a number of centralized, shared research facilities and resources that are utilized by the 11 faculty members, associated staff and students.

RESEARCH

LSBM basic research focuses on learning how genes translate into cell behaviors, not just finding out what genes are involved. They have a strong interest in the mechanisms of cancer and arteriosclerosis, among other things, and have established a number of cell-based model systems for the study of these diseases. They use genomic measurements and genetic and biochemical perturbations to probe for disease-related genes and to study how the gene products function in a disease process in their model cell systems. They emphasize the measurement and analysis of temporal profiles and spatial localization of proteins as part of this work. Dr. Kodama noted that their basic science research on mechanisms of cell function is not as advanced as their applied antibody research.

LSBM applied research is focused on the development of antibodies for diagnostics and therapeutics. This development and commercialization is a major metric of success for the laboratory, according to Dr. Kodama. Specifically, they have a goal of developing 150 new antibodies per year, some of which they will choose to develop for commercialization. To date they have generated several hundred antibodies, and of these they are taking forward five antibodies for diagnostics and two for therapeutics. Their goal for 2008 is to have generated 1,200 antibodies, and be developing 10 of those for therapeutics and 42 for diagnostics.

To carry out this work, they have developed several facilities and resources within the laboratory to support transcriptome analysis, informatics, protein expression, antibody generation, and assays/histology/pathology. The last in particular utilizes collaborations with medical schools and other universities. They discussed in more detail their gene expression database, their protein interaction database compiled by text mining and analysis, and their large-scale generation of antibodies.

Dr. Ihara described the protein interaction database platform that they've developed to help understand mechanisms and select protein targets against which to raise antibodies, in development since 1999. Using the platform they have built pathway maps by literature mining, analyzing more than 1.6 million abstracts in PubMed and extracting more than 700,000 protein-protein interactions through noun-verb-noun combinations. This platform consists of several parts including a dictionary, an index of articles and a way to query gene names. Pathways are generated from pairwise interactions resulting from such a query. Dr. Ihara discussed some of the challenges to this work, including the fact that "too many" interactions make the resulting networks very dense and large, and that a large number of errors occur in the database and pathways in large part because many genes have multiple names. The database is accessible within the Institute now, and a prototype is accessible on the web. They're patenting some aspects of the work. A paper has been submitted for publication, and Dr. Ihara expects to make some parts of the database available publicly once the paper is published.

Professor Aburatani presented their work on a gene expression profile database as infrastructure for the laboratory, and for the public eventually. They're using array-based high-throughput biology of various types, including typing arrays and tiling arrays using Affymetrix chips. The typing arrays are used to find the copy number of chromosomes as well as typing information. With the former they can discriminate which parent is the source of the chromosome as well as copy number. Hence they're using arrays for analyzing mechanisms, not just identifying genes. They're integrating a number of genomic approaches including single nucleotide polymorphism (SNP) typing and ribonucleic acid (RNA) expression. There is a dynamic aspect to their work; for instance some experiments include running transcriptome arrays every 15 minutes for eight hours.

Once they have selected a protein against which they want to make a specific antibody, they express membrane proteins on a baculovirus and inoculate the mouse to raise antibodies against the membrane proteins. Mice do not immunologically recognize the baculovirus, making it a good vector for this work. The cleaved N-terminal protein is also useful for serum detection of protein. A major effort within LSBM is generation of antibodies that can be used for the basic research as well as potential commercialization.

The LSBM works closely with the university's computer-human interface group to visualize data in revealing ways; for instance to visualize the aberrations between patients, and for chromosomal data analysis.

As an application of these resources, cancer is a major focus of the laboratory. They have found a number of cancer-specific genes from microarray applications since 1999. They're currently generating antibodies against certain cancer-associated genes. Liver, stomach and lung cancer are their focus because they're the most common cancers in the Japanese population. In one project, they have generated an antibody against heparan sulfate which they've shown kills liver cancer. It is now in development as an anti-cancer drug.

Another application focus is on nuclear hormone receptor genes. They have expressed 45 of 48 known nuclear hormone receptor genes and raised antibodies against them. They are using the same technologies as above to probe the function of the nuclear hormone receptors and find associated genes.

On the diagnostics, they would like to be able to diagnose a condition from a single drop of blood. They don't expect to find a single marker for a cancer but rather combinations that can form a diagnostic.

Their interest in commercialization of their antibodies as therapeutics and diagnostics requires an emphasis on developing and protecting intellectual property (IP). LSBM collaborates with at least eleven companies, necessitating significant effort on how IP is shared as well. LSBM utilizes their status as a public, academic institution to access resources such as unique reagents that are typically shared among academics, while handling the knowledge gained through their research use carefully so as not to impair IP protection. For instance, Dr. Kodama noted that they would forgo publication and/or patenting a finding in order to keep their comm.-ercialization options open. He also noted that they often work without formal agreements in place so that they can proceed with research. One of the faculty members specializes in medical economics, and Dr. Kodama implied that this faculty member was central to handling IP issues for LSBM.

PERSONNEL

There are about 120 personnel at LSBM, including the following professors:

- Kodama—Director; cholesterol, atherosclerosis
- Ihara—Previously director of research at Life Science Division at Hitachi
- Aburatani—Genomics
- Shibasaki—Membrane traffic and visualization
- Nomura—Previously at a pharmaceutical company, cloned granulocyte colony-stimulating factor (G-CSF)
- Sakai—Previously in Dallas, metabolism and diabetes
- Reid—Membrane microdomains
- Noguchi—Chemistry
- Minami—Formerly at Harvard, vascular biology
- Sakihama—Formerly at Brigham and Women's Hospital in Boston
- Moriguchi—Medical economics
- Hamakubo—Expression methods for membrane proteins

There are about 20–30 graduate students and about 20 postdocs. The rest of the 120 or so personnel are technicians. They don't have undergraduates. While they do wet biology, a number of the researchers have backgrounds in math and physics, which, Dr. Kodama noted, influences their thinking about research approaches. Dr. Kodama noted that a number of people from companies also work at the Institute as part of various collaborations.

TRAINING

The graduate students are educated within the Institute and in other departments, including training in informatics, biology and analytical tools. No continuing education courses are available for outside researchers, although there are many lectures and seminars and people are very interactive within disciplines. A significant aspect of student training is through interaction with people from various disciplines, facilitated by everyone being housed in one building or very nearby. To foster interaction they have a weekly lab meeting with all the graduate students, as well as a two-day annual retreat. After graduation, about a third of their graduate students go to work in industry, a third stay at universities in Japan as postdocs and a third go to the U.S. (presumably as postdocs).

FUNDING

The LSBM budget is about $10 million/year, with the government accounting for about 40% of their funding right now. They get funded through multiple mechanisms:

- Licensing for their patented inventions.
- RCAST has a venture fund, Advanced Science and Technology Enterprise Corporation (ASTEC), which has funded ($10 million to date) a spin-off company, Perseus Proteomics. Some of these funds come back to LSBM for collaborative research.
- Company collaborations. Kowa is a collaborator, providing $4 million to the department from the commercialization of one of their antibodies.
- Funding from the government as part of the "Millennium Project."
- Various government ministries.

APPENDIX D: SUMMARY OF U.S. WORKSHOP, JUNE 4, 2004

INTRODUCTION

For 40 years, biologists have determined phenotype through single gene defects. This extraordinarily powerful approach has been the major contributor to an understanding of how individual genes and proteins function. However, individual gene action appears less likely to provide biologists with an understanding of the behavior of complex biological systems, ranging from individual cells to entire organ systems. Indeed, the assumption underlying systems biology is that phenotype is governed by the behavior of networks. In its essence, systems biology is the development of approaches to the understanding of biological networks and, consequently, to the determination of biological phenotype.

The behavior of networks is too complex to be understood intuitively. Therefore, although a variety of tools may be used to identify the components and connectivity of networks, computational simulation and modeling are widely used to provide that understanding instead. The systems approach is a more effective way to understand input-output behavior of a network, i.e. the relationship of initiating signals to the phenotypic outcome. The models are then modified to account for new experimental results. This iterative process can yield models that may ultimately be queried in ways that are difficult—or even impossible—to accomplish experimentally.

The primary goal of the "Assessment of International Research and Development of Systems Biology" study is to gather information on the status and trends in systems biology from institutions around the world, and to disseminate this information to both government decision makers and the research community. Another goal of this study is to evaluate the state of the art in computational approaches to understand the role of biological networks in determining phenotype. The study also examines the effectiveness of approaches for linking experimental (*in vivo*) parameters with computational (*in silico*) models of network behavior. The study examines issues such as modularity, robustness, motifs, and network topology, as well as the tools that can be used to determine temporal and spatial relationships.

Following this workshop the study panelists conducted site visits to facilities in the United Kingdom, Germany, Belgium, the Netherlands, Switzerland, and Japan to gather information on systems biology and compare the research to that being pursued in the United States. The purposes of the comparative analysis are:

- To identify ideas from overseas programs that are worth exploring in research and development (R&D) programs in the United States
- To clarify opportunities for research and promotion
- To identify opportunities for international and interagency collaboration
- To evaluate the position of foreign research programs relative to those in the United States

SUMMARIES OF EXPERT PRESENTATIONS

Context of the Systems Biotechnology Study

Dr. R. D. Shelton

Dr. Shelton opened the workshop by explaining the context of the systems biology study within the framework of WTEC's mission. Past and current projects, as well as possible future projects, were reviewed. Dr. Shelton pointed out that WTEC has completed more studies of this type than any other U.S. organization, and he discussed the expertise of WTEC staff in this area. He also discussed WTEC's process for organizing and conducting science and technology (S&T) assessments, and explained how WTEC disseminates the results to a broad audience through print and electronic media.

Dr. Shelton outlined the purposes and value of S&T assessments in general. He also reviewed why hosts for WTEC teams are traditionally cooperative, citing the scientific tradition of sharing research results and the benefits to the hosts resulting from technology transfer and professional collaboration.

1. EXPERIMENTAL/DATA TECHNOLOGY

Introduction

Fumiaki Katagiri

Dr. Katagiri discussed the two main trends that he perceives in technology-driven data generation. The first is the establishment of fields of study classed as "-omics," the constituents of which are referred to as "-omes." These fields are very large and consist of single types of data arranged in massively parallel assays—for example, transcriptomes and interactomes. The second trend is the availability of higher spatial and temporal resolutions made possible by higher sensitivity in imaging technology and the use of non-invasive methods capable of tracking a single event. He pointed out that the presentation by panelist Fabio Piano discussed ways of integrating different types of data provided by different technologies.

Over time, the systems biology community is gradually developing a better sense of what can be done with the types of data generated by research. Originally, the reason for generating data was simply because it was possible to do so, but the community needed to discover the following: what types of information could be extracted; what types of data are required for particular purposes; and how these types of experiments could be done properly. Dr. Katagiri pointed out that more data is needed, along with more analysis of that data, to better understand the range of possible applications.

Dr. Katagiri then examined the two predominant ways of looking at data using network modeling techniques, namely top-down and bottom-up. The top-down, or decomposition approach, uses "-omics" phenomenological data to define relationships among biomolecules in the network. The bottom-up, or reconstitution approach, involves quantitative modeling of relatively small, isolated networks. For this approach to be applied successfully, researchers need to know the complete network topology and biochemical information; for example, the concentrations of molecules and any applicable kinetic constraints. A gap is perceived between these two approaches, about which Dr. Katagiri expressed hope that the workshop presentations would suggest ways to narrow and perhaps ultimately close.

Interactome Networks

Marc Vidal

Dr. Vidal emphasized the importance of experimental setups that are suitable for large-scale analyses and that offer versatilities in downstream experimentation. As an example, he summarized his research into open reading frames (ORFs) cloned in a Gateway vector. Dr. Vidal emphasized that data from a large-scale analysis need to be continuously revised and that funding mechanisms should be structured to accommodate such continuous revisions. Rather than insisting that each version be complete before moving on, researchers should instead identify problems and areas of incompletion and proceed to the next version, where solutions to those problems are likely to be found later.

Whereas the traditional biological model has described the impact of the genotype on the phenotype in relationship to the environment, systems biology requires that they be considered as components within in a more complex system that together lead to an output. Relationships are clearly seen between components and outputs, for example in disease, but recently biologists have come to see that between the genome and the phenome are more complex networks, often illustrated as wiring diagrams (see Figure D.1). Dr. Vidal asserted that the relative failure of molecular therapeutics has been due in part to a perception of connections between genome and phenome as a linear path.

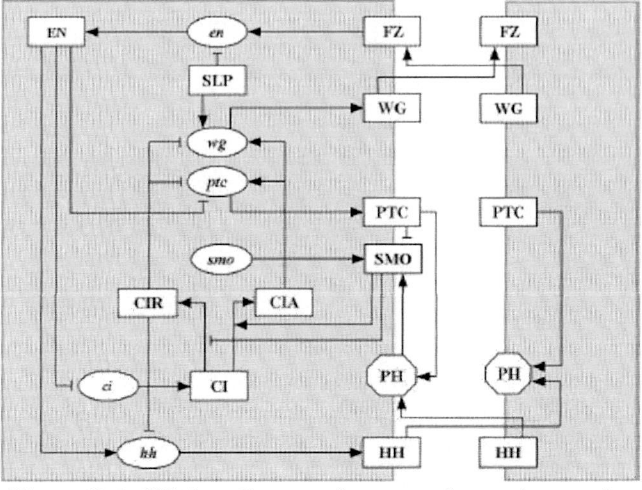

Figure D.1. Wiring diagram for a regulatory interaction.

Reviewing his own work with mapping the genomes of *C. elegans*, Dr. Vidal emphasized that, in order to generate a complete genomic atlas consisting of individual maps placed on top of each other, researchers must first learn how to mass produce the parts from the parts list in adaptable, efficient, and standardized ways. Furthermore, researchers must share resources among themselves, which is a persuasive reason for ensuring that not all research of this type is performed in the private sector.

Dr. Vidal discussed how repeated analysis may also reduce the level of false negatives. Although low false positive rates are often emphasized in large-scale analyses, low false negative rates are also important for the purpose of modeling studies, in which a large amount of information is required. He concluded by stressing the need for more precise perturbation assays, noting that in his research he prefers to perturb edges rather than nodes, though both methods pursued in parallel will yield more comprehensive results.

Phenomics: *in Vivo* Functional Analysis of the Genome

Fabio Piano

Dr. Piano's presentation emphasized the importance of collecting, analyzing, and disseminating complex data on the behavior of genomes *in vivo*. Biologists are faced with the need to forgo the traditional "nodes and edges" approach and ask instead what is it that genes actually *do*. Dr. Piano outlined three challenges facing biologists: 1) decades of research into *C. elegans* has identified only a small percentage of the genome's *in vivo* functions; 2) biologists need to make available new ways of presenting their data, particularly through web-based, searchable databases; and 3) the ability to apply research predictively with complex dynamic networks. Dr. Piano discussed each of these challenges in detail.

In reverse genetic experiments using ribonucleic acid interference (RNAi) at New York University, Dr. Piano and his colleagues found that much effort is required to determine a good measurement of a false negative or a false positive (see Figure D.2). Classic mutations that have been thought to be nulls for the past twenty years, for example, have turned out not to be nulls after all. This is a result of not understanding the "noise" in phenotypic data.

Due to practical limitations inherent in large-scale testing, predictions are necessary and very important. Better spatiotemporal information is needed from the data that is obtained from large-scale analyses. Some of the data will only represent the situation that *could* happen, which could be different from the *actual* situation under certain conditions. Therefore, when the data are to be used in modeling studies, condition-specific information is essential.

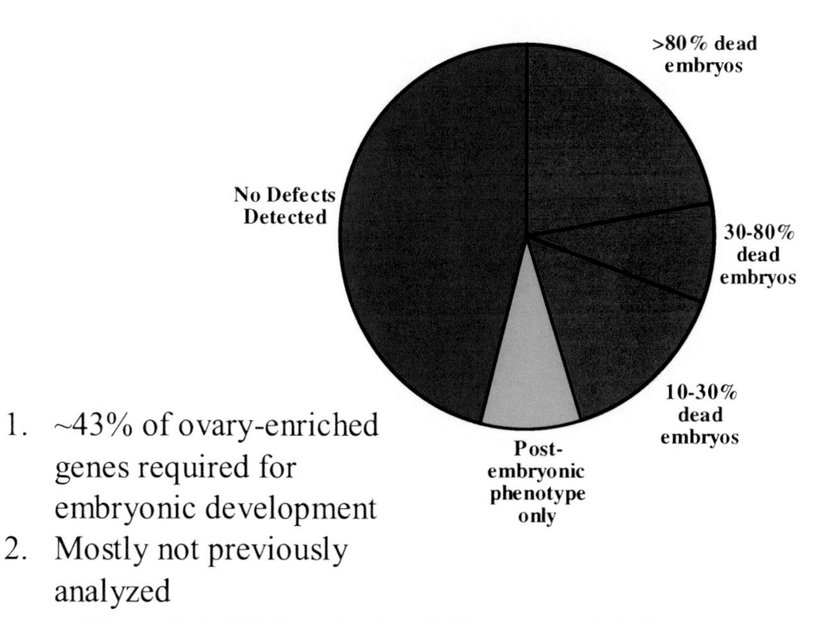

Figure D.2. RNAi analysis of 750 ovary-enriched genes.

Dr. Piano emphasized the importance of recording phenotypes in multiple aspects, such as sequential time points, to identify the multiple roles of particular genes. For example, rather than assume that the first effect seen downstream of a gene removal is a direct effect that in turn identifies the gene's function, researchers are now able to assess complex patterns of defects by searching and viewing all relevant video footage in the RNAi database. The database allows primary screens to be performed on the computer rather than in the laboratory, thereby improving predictability (see Figure D.3).

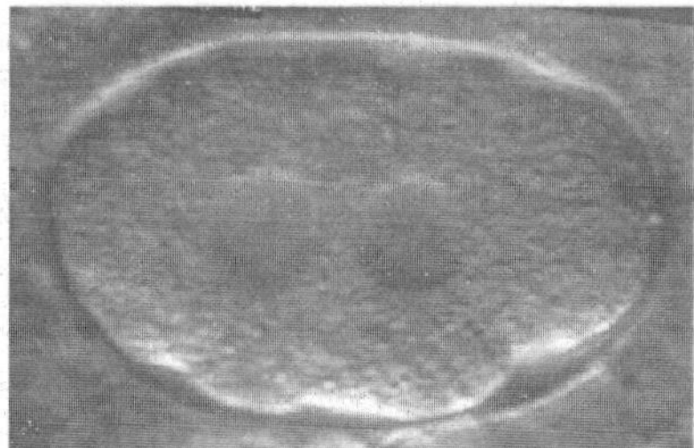

C36B1.5

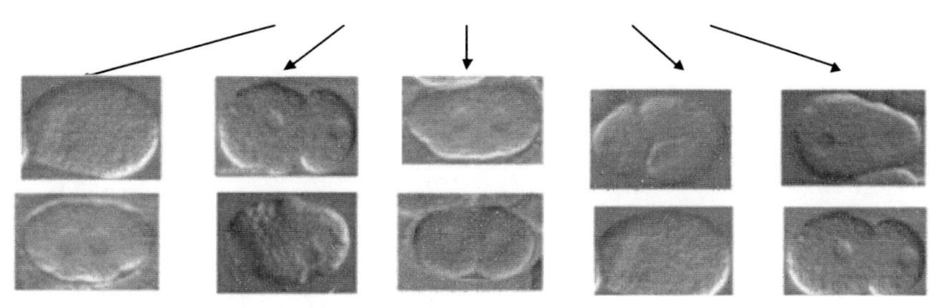

Figure D.3. Searching the database for similar phenotypes.

2. DATA ANALYSIS AND SYSTEM INFERENCE

Systems biology is an approach that is being applied successfully to address many problems in biology. This session provides an overview of various systems biology approaches that are being used to integrate biological data from diverse sources and making such integration useful in pharmaceutical, environmental, and basic biomedical research. Specific examples in industry and government and academic labs are described. In addition, the logistics and infrastructure required to successfully implement systems biology to make progress in understanding complex biological systems are discussed.—*Cindy Stokes*

Learning Regulatory Networks from High-Throughput Genomic Data

Daphne Koller

Dr. Koller discussed the application of probabilistic graphical computational models to the analysis of heterogeneous data to obtain meaningful biological information. She began by reviewing the range of applications of computational modeling. Better data sets and better learning algorithms will, over time, result in more robust fine-grained models that will provide more insight into biological systems. Questions that arise with regard to modeling are representation—what can be modeled—and learnability—what can be learned.

Probabilistic graphical models represent a biological system as a set of multiple interacting variables. They allow researchers to capture the inherent non-determinism in a system and compensate for noise in the available assays. Direct and indirect variable interactions can be modeled, allowing for a wide range of model complexities.

Dr. Koller provided an overview of the applications for probabilistic modeling, noting their usefulness for developing sequence, evolutionary, and *Cis*-regulatory models, identifying gene regulation from gene expression data, and mapping physical interaction networks. Gene regulation models were discussed in more detail. One of the first applications of this model was a Bayesian network analysis in which the variables were the expression levels of genes, which are seen to be a surrogate of gene activity (see Figure D.4).

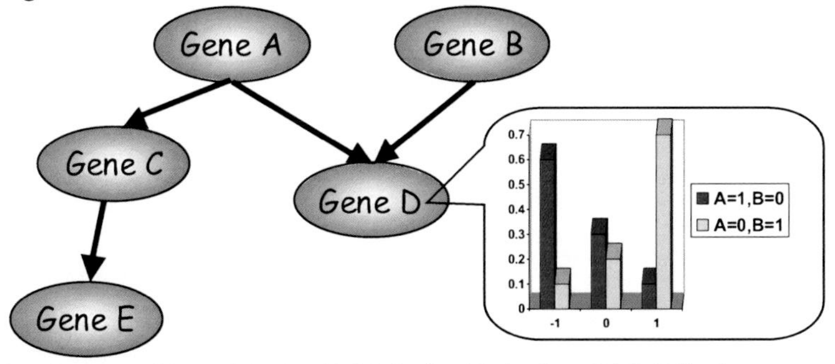

$$P(A,B,C,D,E) = P(A)P(B)P(C|A)P(D|A,B)P(E|C)$$

- Each node is a random variable
  - Indicating mRNA expression level of some gene
- Arrows indicate direct probabilistic interaction

Figure D.4. Bayesian network analysis of gene expression levels.

Dr. Koller discussed the construction of a statistical modeling language that allows direct representation of heterogeneous biological entities such as genes, proteins, or transcription factors, and the mechanisms by which they interact. The language was designed to encompass object classes, properties, and interactions. The learning algorithm uses raw data to recover gene assignments to modules and thereby recover the regulatory program that defines how the genes respond to certain configurations of regulators.

The algorithm was tested in the wet-lab on high-throughput genomic data sources. For yeast stress data, 46 out of 50 modules were found to be functionally coherent, while 30 out of 50 regulatory relationships occurred as predicted (see Figure D.5). A subsequent experiment using three modules found that all three regulated computationally predicted genes. The probabilistic framework allows these different data sources to be integrated easily and coherently, allowing the limited and noisy views provided by different assays to be combined. The use of a high-level language that directly encodes biological entities and processes allows biological insights to be derived from the learned model.

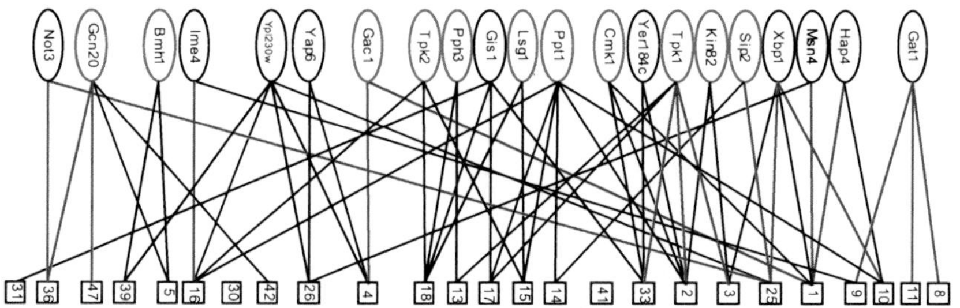

- Are the module genes functionally coherent? **46/50**

- Are some module genes known targets of the predicted regulators? **30/50**

Figure D.5. Global module map of an algorithm test.

Applications of this framework to the analysis of genomic data include understanding the process by which DNA motifs in the promoter regions of genes play a role in controlling their activity, finding regulatory modules and their actual regulator genes directly from gene expression data, and combining gene expression profiles from several organisms for a more robust prediction of gene function and regulatory pathways. Because probabilistic models are a unified framework, they allow for the introduction of additional information.

Dr. Koller concluded by examining possible future applications of probabilistic modeling for the development of better mechanistic models of gene regulation, reconstructing interaction networks, multi-species analysis, and for the testing of new data types ranging from phenotype data to chromatin structure.

Identifiability Issues in Modeling Regulatory Networks

Frank Doyle

Dr. Doyle began his presentation by pointing out that his approach to the topic is from the perspective of a systems engineer. He framed his presentation in terms of two questions: how far can researchers get using data, and can models identify the approaches that offer a maximization of information content (see Figure D.6)? To obtain more data for more complex systems, the most efficient experiments must be identified. Transcription and signal transduction algorithmic models were reviewed.

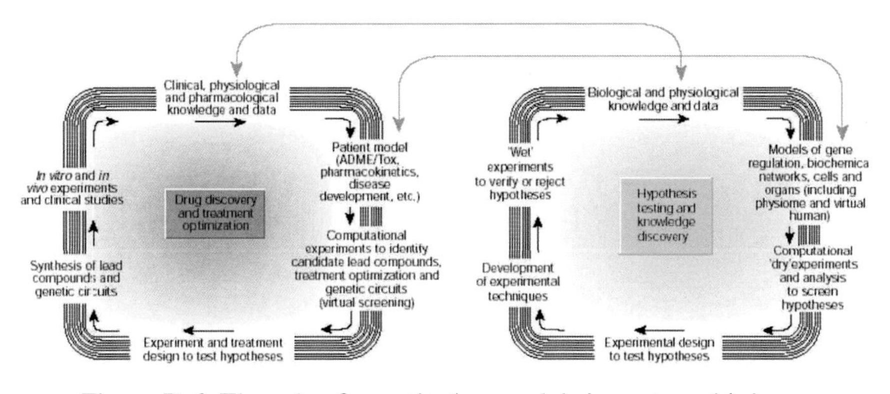

Figure D.6. The role of quantitative models in systems biology.

The application of models to linear and low-dimensional systems, while mature in systems engineering, has little or no applicability to biological systems. Dr. Doyle highlighted the key elements of system identification, namely model structure selection, coefficient regression, and model validation (the concept of model *invalidation* being preferred). The questions to be answered include whether the identification procedure will yield a unique parameter set; the degree of correlation between the model and the true system; and the potential for model discrimination through experimental conditions. Dr. Doyle cautioned that while global results can be derived from linear models, nonlinear models provide only locally applicable approximations.

The Fisher Information Matrix can be applied as a measure of information content quality. The matrix plots the bounds of uncertaintly as a hyperelliptical plot on a two-dimensional graph, allowing researchers to select measurements that enrich the desired orientation (see Figures D.7 and D.8). This approach offers additional insight into system sensitivity and robustness.

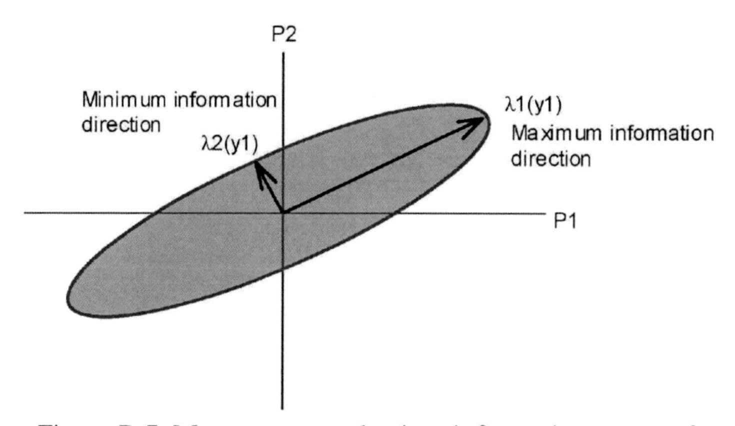

Figure D.7. Measurement selection: information content for the first measurement.

Dr. Doyle identified several qualitative and quantitative design factors that affect the index of identifiability that is derived using this approach. Of the qualitative aspects, inputs (e.g. ligand, environmental, and knock-outs) is the most important; measurement and quantitative factors such as the perturbation richness, experiment duration, sampling protocol, and the number of both samples and cells per sample were seen to be issues of data quality. Dr. Doyle provided overviews of studies that differed qualitatively and quantitatively. These studies all suggest that the asymptote is in the region of 75–80%. The experimental consequences of model mismatch were also reviewed.

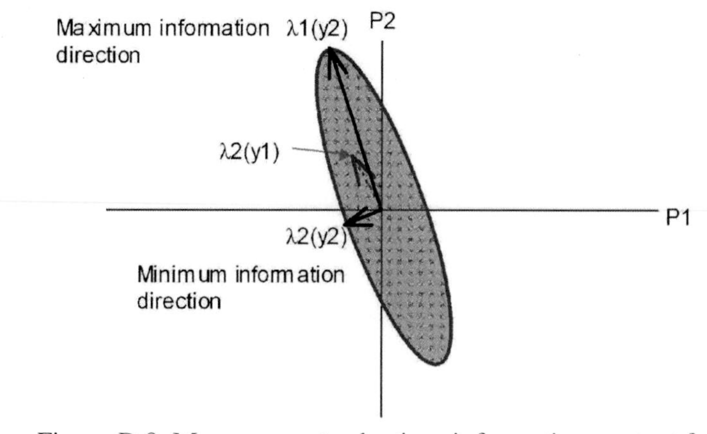

Figure D.8. Measurement selection: information content for the second measurement. The second measurement is chosen such that there is maximum information in the direction of prior minimum information.

In conclusion, Dr. Doyle stressed the need for increasingly formal approaches to identifiability, experiment design, and protocol design in systems biology. Identifiability is a bottom-up approach that can be integrated with other top-down approaches currently being pursued. Because the approach can be used to analyze robustness, it allows researchers to analyze design principles more completely. Dr. Doyle stressed that what is needed is not simply more data, but a concurrent rethinking of protocol.

3. NETWORK ORGANIZATION

One facet of biomolecular networks in cells concerns their organization, which may be taken to mean at least some of the following features: a) topology among molecular nodes; b) relationships by which certain nodes influence others; and c) their operational characteristics. Any or all of these features might be perceived as representing "design" properties in human engineering terms. Thus, a set of questions that may be posed for address on this topic include: 1) Can "structure/function" relationships be found in network topologies? 2) Can analogies to human-made networks be discerned, whether for intuitive understanding or for identifying intervention strategies? 3) Can "lessons" informing design or development of novel human-made networks be elucidated?—*Douglas Lauffenburger*

Network Navigation: From Data Collection to Visualization to Perturbation

Mike Tyers

Dr. Tyers discussed network topologies, emphasizing the dramatic distinctions between physical and functional interactions. Yeast offers a systematic biology toolkit for a wide range of research, including DNA microarrays, protein-DNA interactions, localization, synthetic lethal genetic networks, and protein arrays (see Figure D.9). While his laboratory does not do systems biology, they do undertake systematic approaches to problems in which they are interested.

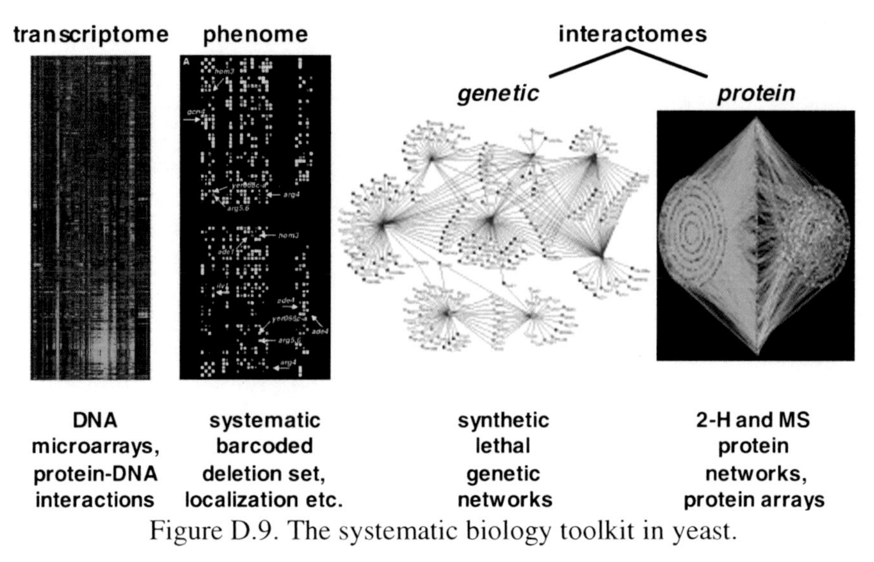

Figure D.9. The systematic biology toolkit in yeast.

Dr. Tyers pointed out that what is lacking from the networks is the dynamic regulations typically controlled by various post-translational modifications. The platform for identifying protein interactions is out of date and is being updated. To build and curate datasets, laboratories develop locally applicable informatics systems that are not compatible with systems developed elsewhere. Dr. Tyers emphasized that exchangeable formats are a minimum requirement for significant future progress.

Because networks are dynamic, static pictures must be compiled on top of each other to understand how networks change over time and condition. An effective proxy is the ability to monitor post-translational modifications. This approach promises a host of practical applications for drug development. Many unique photopeptides can be identified and compared, and differential regulation is possible to find. Results have shown that approximately 80% of the sites do not match known kinase consensus sequences. Molecular biology and biochemicals view of how kinase works is not reflected on a protein level.

Dr. Tyers then discussed his laboratory's systematic research program into synthetic lethal genetic interaction networks. These networks can be understood as parallel pathways that intersect on an essential function, resulting in lethality. The laboratory's high-throughput method yields an average of 30 to 35 interactions per gene, whereas earlier methods were able to obtain only three or four. The high density of genetic interactions has implications for polygenic disorders and drug discovery (see Figure D.10).

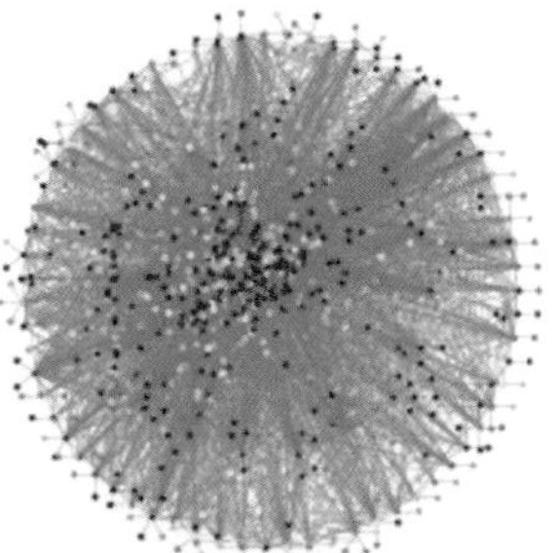

Figure D.10. Towards a global genetic interaction network in yeast.

Systematic perturbation, or chemical synthetic lethality, was discussed in terms of lethal screens in human cell lines. While the ideal of node-specific drugs is a long-term goal, a reasonable near-term goal is screening for drugs that are specific to certain genetic contexts in sub-networks. The goal is to not only build chemical lethality networks, but also to find out how chemicals interact with each other.

Dr. Tyers reviewed several specific concerns related to these approaches. Critical data is missing from high-throughput data. Researchers are having difficulty sifting irrelevant details from the overwhelming data. Systematic testing of networks presents a fundamental challenge. It is also essential to differentiate between population effects and single cell effects, to identify the relevance of readouts, and to identify non-obvious evolved properties. He concluded by pointing out that existing models are incomplete and hundreds of genes remain unaccounted for.

Robustness as a Design Principle of Biological Networks

Naama Barkai

Dr. Barkai's discussion of operational characteristics emphasized the utility of robustness as a design principle for biomolecular networks. It is not currently realistic to model full cells or cellular networks, e.g. metabolic networks. Instead, bioinformatics can be used to identify design principles and to elaborate small networks. The most important challenges that Dr. Barkai sees is to develop explanatory models that elucidate core mechanisms and eliminate incompatible ones while predicting crucial molecular features. Prediction must distinguish between model behavior and crucial molecular features or crucial feedback points while attempting to discern general design principles that underlie network functions.

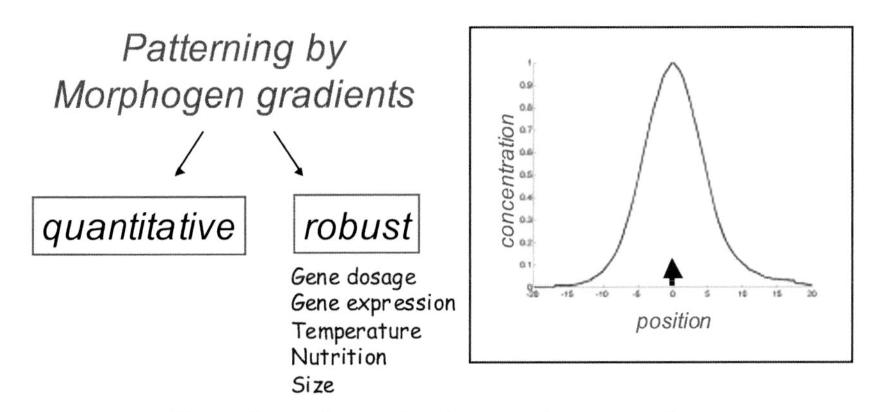

Figure D.11. Patterning by morphogen gradients.

The system to be studied should be determined on grounds of biological issues, not the availability of the system to the researcher. If the research interest is robustness or specificity, the selected system should be well-characterized and accessible to quantitative experiments. Dr. Barkai's own case study focused on elucidating robustness as determined by patterning morphogen gradients (see Figure D.11). While robustness is chiefly thought of as a design principle, it can also be used as a tool for distinguishing between plausible models and a means by which the mechanism that generates robustness can be discovered.

When comparing a canonical linear model of morphogen gradients with a nonlinear model, the question arises as to which model is more compatible with the observed data. The "liberal approach" is to start with a permissive model that allows interruptions and is compatible with different mechanisms. The patterning mechanism as a gradient of inhibitor is found to be not robust, while the mechanism as a gradient of activators is found to be robust.

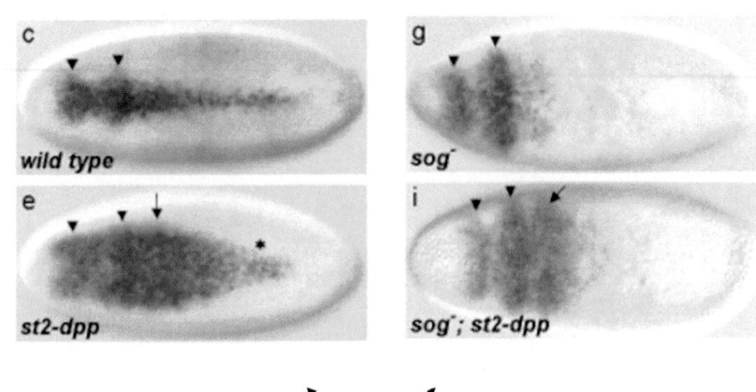

Figure D.12. Testing morphogens in zygotes.

Dr. Barkai concluded by reviewing case studies in which quantitative data were obtained by testing morphogens in zygotes in which both steady state and non-steady state models were applied (see Figure D.12). Perturbation of the profile was found to be lower in the non-steady state.

4. SYSTEMS MODELING

Biochemical Computation of a Mammalian Cell

Ravi Iyengar

Dr. Iyengar discussed methods for developing biochemical models of regulatory networks as a basis for detailed models of a mammalian cell in terms of cellular components and their interactions. By thinking of a cell as a complex chemical plant with master controller systems with semi-autonomous sub-controllers and a set of interconnected reaction vessels, new interactions can be tracked (see Figure D.13). Networks can be analyzed qualitatively, by understanding regulatory motifs, and quantitatively, using physical chemistry representation of reactions in deterministic and stochastic models incorporating a representation of spatial dynamics.

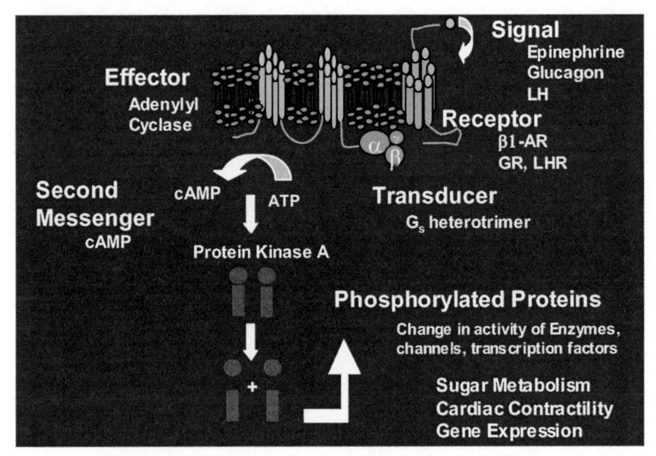

Figure D.13. A linear signaling pathway.

Qualitative analysis of a component-based representation of the neuron has yielded results that indicate that short-range regulatory motifs are favored. The extent of the connectivity propagation varies for different signaling pathways, whereas highly connected nodes introduce more positive feed-forward and feedback loops. The presence of these regulatory motifs allows for signal persistence at potentially many components. To construct large-scale maps, complete biochemical specifications for direct interactions and pathways are needed.

Quantitative analysis of the network process follows the collection of the above data. Using an oversimplified model, the bistable behavior of the feedback loop was found to indicate a highly flexible regulatory switch (see Figure D.14). Because blotting introduced spatiotemporal asymmetry, the results are not representative of complete cell behavior.

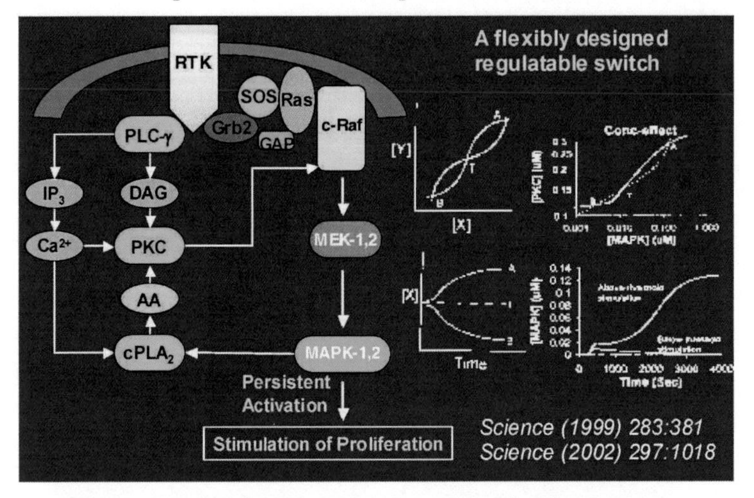

Figure D.14. Bistable behavior of the feedback loop.

Dr. Iyengar then discussed the need for a quantitative data management database system that would be capable of identifying cellular concentrations, interaction constants, enzymatic activity, subceullar activity, flux, and the rates of regulated movement. Dr. Iyengar then went on to discuss spatial representation of signaling networks governing chemical reactions. They provide dynamic regulation of boundary and domain characteristics, particularly anisotropy of coupled chemical reactions and the role of isoform diversity in spatial and temporal dynamics. A case study of the specific functions of PD4E4D, which is very well known, was discussed.

A simulation conducted using a model of a hippocampal neuron generated interesting spatial dynamic characteristics (see Figure D.15). Initial conclusions from this experiment suggest that cellular geometry may help create dynamic microdomains. Boundary dynamics can be defined by negative regulators, while anchoring of the negative regulators contributes to boundary sharpening. Further research is needed to determine local concentrations of native components. A methodology for storing information on local concentrations needs to be developed.

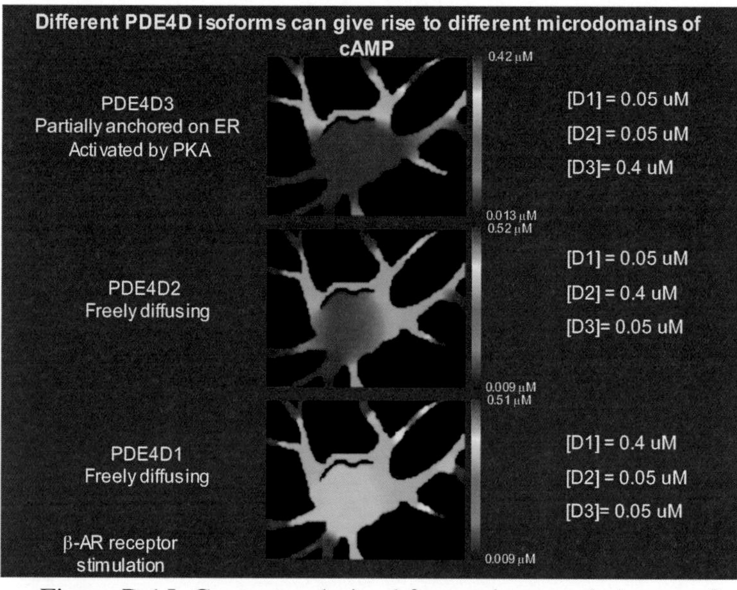

Figure D.15. Geometry derived from microscopic image of hippocampal neuron.

Dr. Iyengar concluded by reviewing a moving-boundary problem case study involving the accurate measurement of the spreading of mouse fibroblast cells as a function of time. While the binding of protein concentrations to the membrane was such that the signaling reactions are reasonably deterministic the diffusion-limited regulation of branching were found to be entirely stochastic. In summary, Dr. Iyengar discussed several experimental and theoretical needs for the detailed analysis of deep large-scale network models.

Reverse Engineering Developmental Genetic Regulatory Networks

Hamid Bolouri

Dr. Bolouri discussed the use of computer software to reverse engineer the developmental genetic regulatory networks in the cells of sea urchins, yeast, and mammalian cells. He posed the following questions to be considered at the outset of a computer modeling process:

- What data is most helpful?
- What are the issues in model verification and validation?
- Why are certain models considered successful and others not?
- What is really the role of modeling in the future of biology?

Dr. Bolouri's presentation emphasized that modeling is useful for developing an understanding of processes, but data, once obtained, speaks for itself. Different systems offer different opportunities; for example, between a sea urchin network, a macrophage pathway and a simple yeast pathway. The process by which a sea urchin cell differentiates can be visualized using software that is available on the web. Perturbation experiments were conducted to describe asymmetry in sea urchin eggs (see Figure D.16). All regulatory interactions have been characterized and the model is considered to be true (see Figure D.17).

Perturbation: Morpholino antisense

Effect: Prevents translation of mRNA

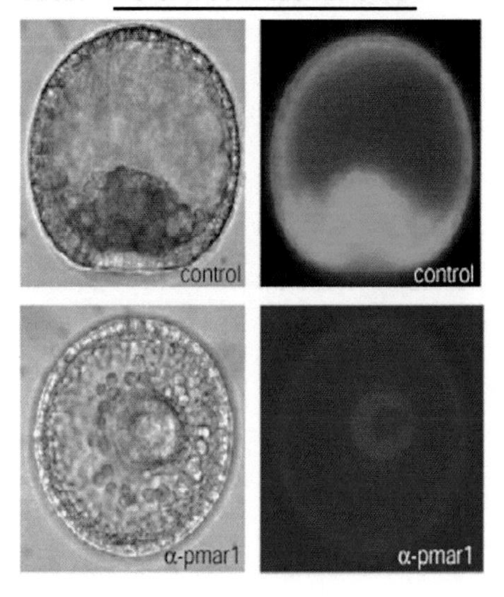

Perturbation: Cadherin mRNA injection

Effect: Blocks activation of Wnt/Tcf signalling pathway

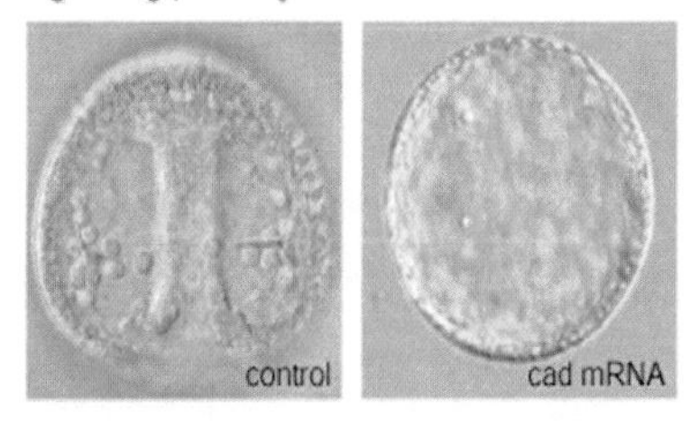

Perturbation: Negative Notch mRNA injection

Effect: Blocks Notch signalling pathway

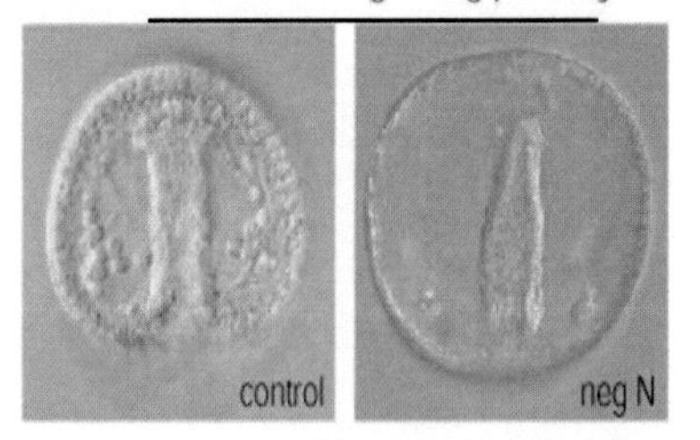

Perturbation: Engrailed repressor domain fusion

Effect: Converts transcription factor into dominant rerpressor for all target genes

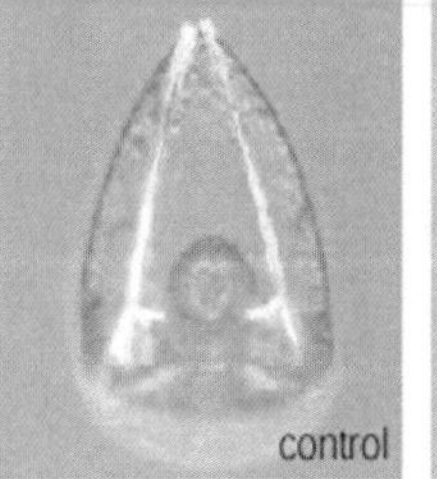

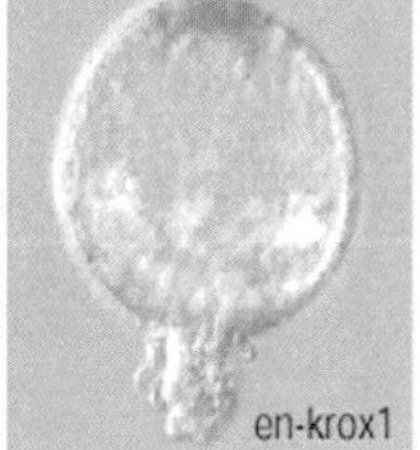

Figure D.16. Network perturbation methods.

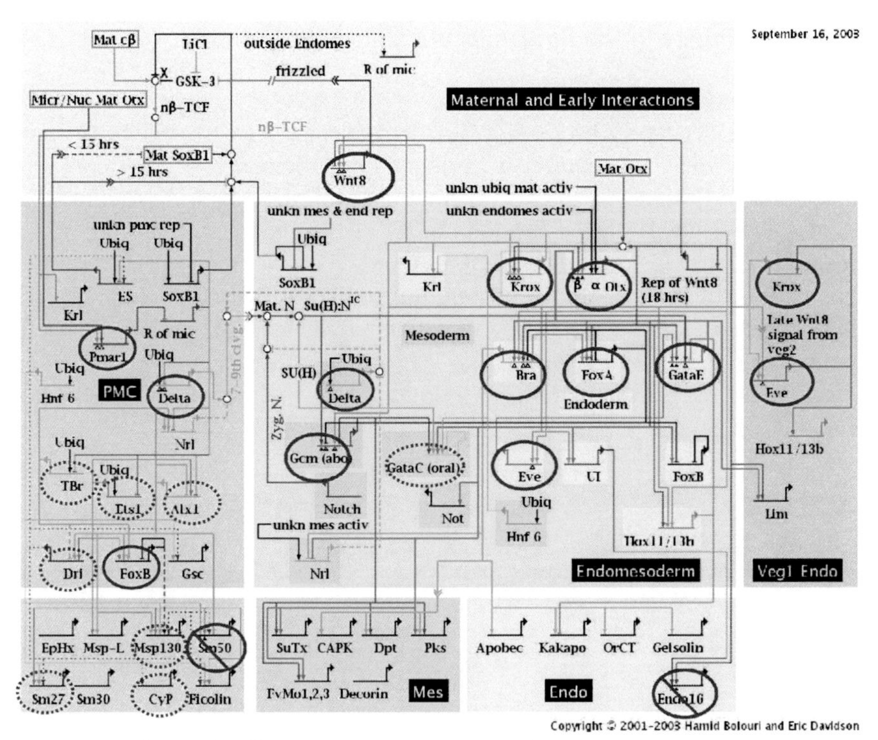

Figure D.17. Endomesoderm specifications to 24 hours.

The sea urchin data showed that genes took longer to reach steady state than previously thought, which has been confirmed by subsequent research. Dr. Bolouri's lab has developed parameterized gene models for rapid construction of networks. Comparing the data with a mammalian system, all interactions are unidirectional and feedbacks to signaling networks were discovered.

5. APPLICATIONS

Applications of systems biology derive from predictive models based on extensive and comprehensive data collections. The models must be capable of predicting the outcomes of new experiments: to accurately predict the result of natural or externally introduced changes in the system, such as alterations to the normal state in disease, perturbation by drug intervention, or genetic variation (mutations). The general ability to model systems in quantitative detail provides the basis for biological engineering, perhaps the most important downstream application of systems biology.—*Chris Sanders*

Systems Biology in the Pharmaceutical and Biotechnology Industries

Cynthia L. Stokes

The drug discovery process is slow and expensive, with recent estimates of the time and cost required to bring a drug to market at fourteen years and $800 million. One major contributor to the increasing burden of drug development has been the limited ability of the industry to predict human response to therapeutic modulation of molecular drug targets. That is, there has been little success in understanding the complex, nonlinear networks that lie between a potential drug target and an efficacious endpoint in a patient population (see Figure D.18). While animal models have helped address this issue, species differences have greatly limited their predictive capability.

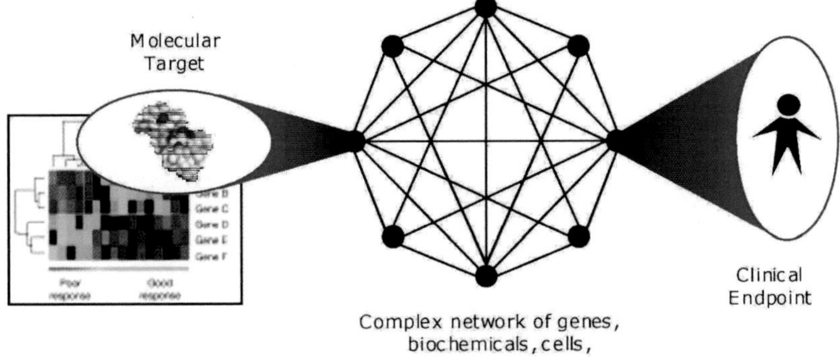

Figure D.18. Disconnect of molecular targets from clinical endpoints is a major problem. Need better ways to predict clinical effects of modulating a target and evaluating potential therapeutic compounds throughout the process.

Systems biology is now advocated as a means to give much-needed systems context to biological and physiological data (i.e. "global data"—genomic, proteomic, metabolomic, clinical, etc.). Systems biology is not defined by any particular technology or tool, but rather it is an *approach* to understanding biological systems and, from a pharmaceutical industry perspective, to making sound decisions necessary to advance targets, leads, compounds, and drugs through the discovery and development pipeline. While maintaining the hypothesis-driven tradition of biological research, systems biology differs from recent reductionist approaches by integrating global data through an iterative process that produces a quantitative and dynamic understanding of human biology and disease.

With a focus on understanding, quantitatively and dynamically, human pathophysiology, systems biology holds the promise for deciphering the enormous amounts of data being driven out of the human genome via today's high-throughput technologies. "Systems Biology" companies, typically using one or more core technologies, are gathering, organizing, and integrating biological data from various sources into "systems" models of human physiology and pathophysiology (see Figure D.19). These models are then applied to a variety of research and development issues related to target identification and validation; lead identification and optimization; compound selection; biomarker identification; and clinical trial design and optimization. Systems biology companies span both wet and dry research, both *in vitro* and *in vivo* experimentation, both animal and human clinical protocols, and include, for example: high-throughput molecular biology data acquisition methods, data-mining algorithms, mathematical modeling, ontologies, and text analysis.

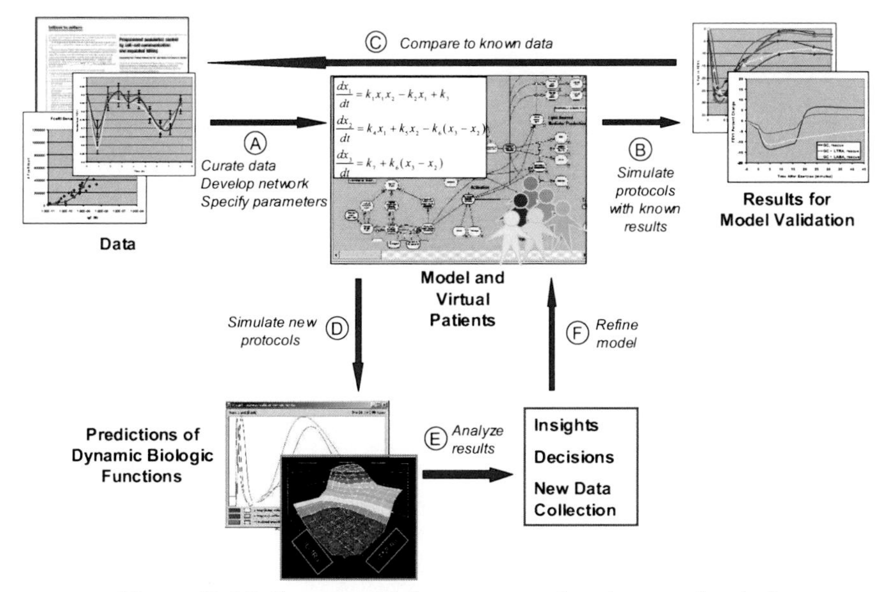

Figure D.19. Systems biology approach using mechanistic modeling as the core technology.

Dr. Stokes discussed examples of the use of systems biology approaches in the pharmaceuticals industry, such as the prospective prediction of exercise-induced asthma treatment. When modeled on a proprietary software platform and compared to clinical data, the results isolated PDE4 as a target for asthma treatment (see Figure D.20). Significant contributions to the development of new therapeutics have been made including the identification, evaluation, and prioritization of drug targets and lead compounds; biomarker discovery; and clinical trial design. Dr. Stokes' presentation illustrated how systems biology approaches are being utilized fruitfully to bridge the gap between molecular targets and clinical endpoints and thereby provide valuable understanding for drug development activities and decisions.

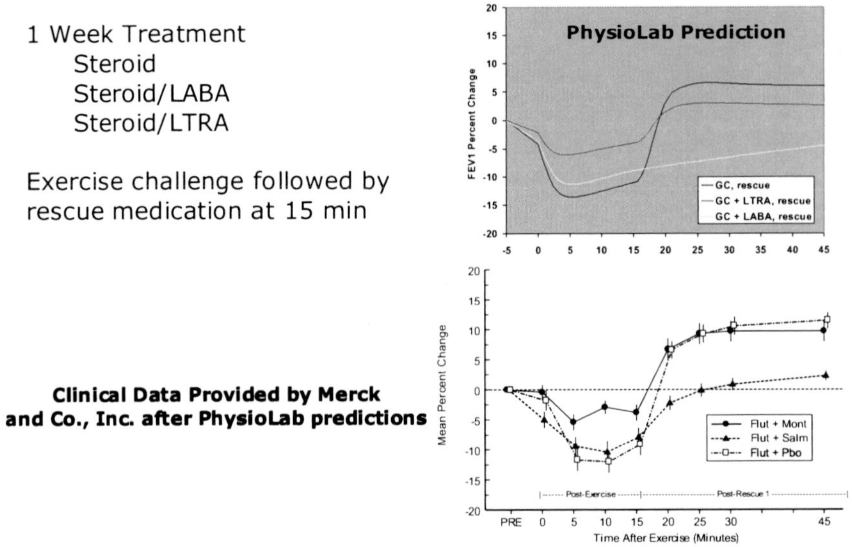
1 Week Treatment
Steroid
Steroid/LABA
Steroid/LTRA
Exercise challenge followed by rescue medication at 15 min
PhysioLab Prediction
FEV1 Percent Change
GC, rescue
GC + LTRA, rescue
GC + LABA, rescue
Clinical Data Provided by Merck and Co., Inc. after PhysioLab predictions
Mean Percent Change
Flut + Mont
Flut + Salm
Flut + Pbo
Post-Exercise
Post-Rescue 1
Time After Exercise (Minutes)

Figure D.20. Prospective prediction of exercise-induced asthma treatment.

Implementing a Systems Approach to Understand Cellular Networks

Steven Wiley

Dr. Wiley described efforts to engineer an efficient pipeline for data collection, including automation and the integration of experimental knowledge from diverse technologies. He illustrated the power of such data in modeling pathways involving the regulation of gene expression as a key to understanding cellular control processes.

To successfully model a complex biological system, the following information must be known:

- What parts are being made?
- What is the regulatory network structure?
- Where are the proteins located in the cell?
- What is their scalar quantity?
- How do they interact with their partners?

Cells viewed here are input-output systems in which a stimulus (independent variable) is applied to a defined system and the resulting dependent variables are observed. The relationship between the input and the output provides insight into the workings of the system. Biologists are now realizing that the context of the cell is an input variable; this is a challenge because genomic and structural data are context-independent. Therefore, the need to define the composition of living systems is the factor that is driving a wide range of analytical technologies such as genomics, proteomics, metabanomics, expression profiling, and imaging. All of these technologies seek to rigorously define the composition of living systems.

Dr. Wiley discussed his lab's effort to conduct global quantitative proteome measurements (identification and quantification) using accurate mass and time tags. High-throughput proteomics is conducted using ultra high-pressure liquid chromatography and capillary electrophoresis columns. Resolutions of approximately 100,000 peptides/run have been achieved, with identification in the range of about 1,400–2,000 proteins/run. The effort to understand the influence of cell context is driving experimental and computational biology. The context must be understood before experiments can be undertaken.

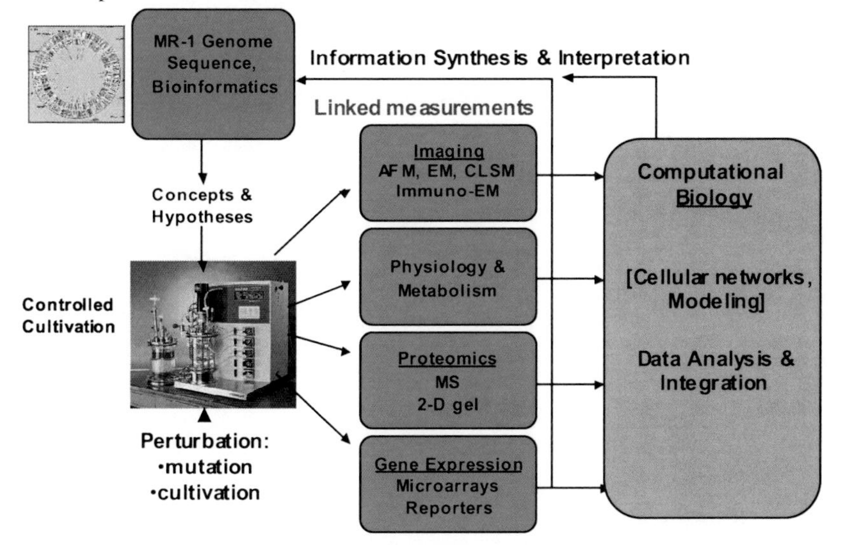

Figure D.21. Systems approach to *Shewanella* biology.

Dr. Wiley then briefly reviewed several case studies in which context-dependent analytical technologies were used. In the DoE's Genomes to Life program, *Shewanella* and other bacteria are bred to convert soluble uranium to insoluble uranium and thereby immobilize it. All samples are generated by a single facility under controlled conditions, which is essential to ensure standardization, reproducibility, and provide an explicit context (see Figure D.21). In a short period of time, the pathways have been mapped and organisms have been reconstructed with the desired characteristics. The automated isolation and analysis of protein complexes using mass spectrometry has significantly reduced noise and provided reproducible data.

The second example discussed factors controlling the regulation of microbial communities. DoE is interested in both intracellular and extracellular networks with the long-term goal of understanding how multi-organism networks in communities operate and are regulated at a molecular level. Structured anaerobic communities are favored because they exhibit tractable levels of genetic complexity; are stable associations that provide for replication and modeling; generate sufficient biomass for analysis including non-destructive imaging; and have sufficiently high levels of activity for perturbation/response studies (see Figure D.22).

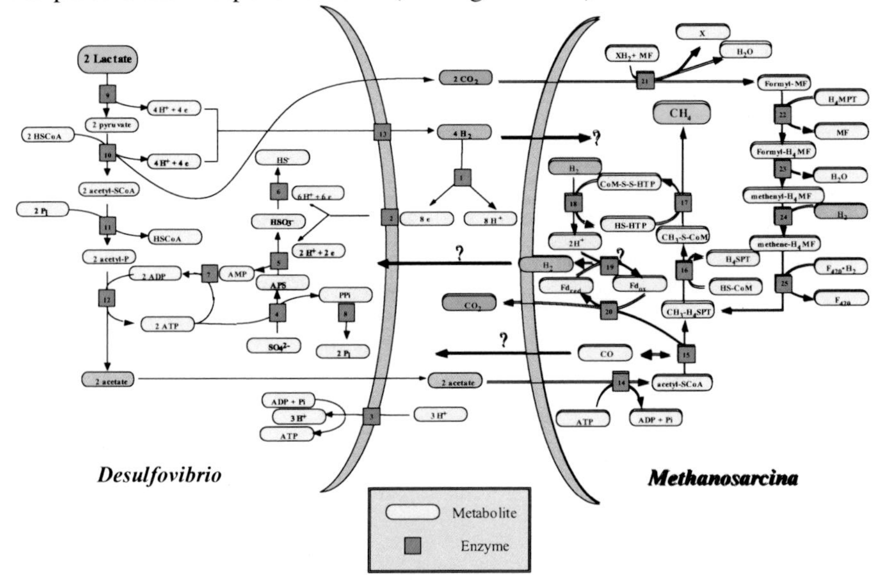

Figure D.22. Coupled metabolic pathways.

In conclusion, cell responses are seen as multiphasic, and that different classes of stimulants are processed at characteristic time scales. Furthermore, processing nodes within cells are seen to be spatially segregated and cells respond independently depending on their specific contexts. Responses generally induce a reprogramming of the cell machinery. To create cell simulations, this information must be abstracted to create a reference model that can then be modified. Several obstacles prevent or hinder the utilization of available data plus several technical and social challenges that must be addressed and overcome for systems biology to continue its progress.

APPENDIX E. GLOSSARY

ADME/Tox	Absorption, distribution, metabolism, and excretion/toxicology
AfCS	Alliance for Cell Signaling
AGRIKOLA	*Arabidopsis* Genomic RNAi Knock-out Line Analysis (Europe)
AIST	National Institutes of Advanced Industrial Science and Technology (Japan)
ASTEC	Advanced Science and Technology Enterprise Corporation (Japan) at the University of Tokyo
ATP	Adenosine triphosphate
AV node	Atrio-ventricular node
BBSRC	Biotechnology and Biological Sciences Research Council (U.K.)
BCT	Bacteria chemotaxis
Beacon projects	Scientific projects launched by the Department of Trade and Industry BioScience Unit in 2002 (U.K.)
BioMS	Modeling and simulation in the biosciences (Germany)
BL-SOM	Batch-learning self-organizing map
BMBF	Federal Ministry of Education and Research (Germany)
CA	Coordination Action (Europe)
CAGE	Compendium of *Arabidopsis* gene expression (Europe)
CAM	Crassulacean acid metabolism
CASP	Critical Assessment of Techniques for Protein Structure Prediction
CATMA	Complete *Arabidopsis* transcriptome microarray (Europe)
CBRC	Computational Biology Research Center (Japan)

CDB	RIKEN Center for Developmental Biology (Japan)
CDKs	Cyclin dependent kinases
CDPs	Cell decision processes
CHGC	Chinese National Human Genome Center (China) in Shanghai
cDNA	Complementary deoxyribonucleic acid
CDP	Cell Decision Processes (CDP) Center at MIT (U.S.)
CellML	Cell Markup Language
CE/MS	Capillary electrophoresis mass spectrometry
ChIP	Chromatin immunoprecipitation technique
CIB-DDBJ	Center for Information Biology and DNA Data Bank of Japan
CMB	Centre for Mathematical Biology (U.K.)
CMISS	Continuum mechanics, image analysis, signal processing and system identification
CoMPLEX	Centre for Mathematics in the Life Sciences and Experimental Biology (U.K.)
COPASI	Complex pathway simulator
COR	Canonical correlation analysis
CPU	Central processing unit
CSB	Computational Systems Biology
CSML	Cell System Modeling Language
CUS	Swiss Universities Conference (Switzerland)
DARPA	Defense Advanced Research Projects Agency (U.S.)
DECHEMA	Society for Chemical Engineering and Biotechnology (Germany)
DKFZ	German Cancer Research Center (Germany)
DNA	Deoxyribonucleic acid

DoE	Department of Energy (U.S.)
DTC	Doctoral training center (U.K.)
DTI	Department of Trade and Industry (U.K.)
DTU	Technical University of Denmark (Denmark)
EBI	European Bioinformatics Institute (Europe)
ECG	Electrocardiogram
EcoRV	*E. coli* restriction enzyme
ECTS	European Credit Transfer System
EGF	Epidermal growth factor
EMBL	European Molecular Biology Laboratory (Europe)
EML	European Media Laboratory (Europe)
ENU	N-ethyl-N-nitrosourea
EPSRC	Engineering and Physical Sciences Research Council (U.K.)
ERA	European Research Area
ERATO	Exploratory Research for Advanced Technology (Japan)
ERK	Extracellular-signal Regulated Kinase
ESF	European Science Foundation
ESI	Electrospray ionization
ESRF	European Synchrotron Radiation Facility (Europe)
EU	European Union
EUREKA	A consortia of firms in partnerships with universities and research institutes from 22 European countries that cooperate on scientific projects directed at developing products, processes, and services having a world market potential.
EUSYSBIO	Specific support activity for the coordination of systems biology in Europe
FANTOM	Functional Annotation of Mouse

FLJ	Full-length long cDNA collection of Japan
FNRS	Fond National de la Recherche (Belgium)
FPGAs	Field programmable gate arrays
FORTE-SUITE	A system developed by CBRC to predict tertiary structures of proteins from the sequence information.
FRET sensor	Fluorescence resonance energy transfer sensor
FT-MS	Fourier transformation mass spectrometry
G-CSF	Granulocyte-colony stimulating factor
GA	Genetic algorithm
GATC-PCR	Generalized adaptor-tagged competitive - protein-coupled receptors
GB	Gigabytes
GC-TOF-MS	Gas chromatography time-of-flight mass spectrometry
GFP	Green fluorescent protein
GMO	Genetically modified organism
G.NET	Gene Network Inference Method
GON	Genomic object net
GSC	Genome Sciences Center (Japan) at the RIKEN Yokohama Institute
G-SCF	Granulocyte colony-stimulating factor
GTL	Department of Energy's Genome to Life project (U.S.)
HCA	Hierarchical Control Analysis
HIF	Hypoxia inducible factor
HMT	Human metabolome technologies
HOG	High-osmolarity glycerol
Hox gene	A gene that contains a homeobox, a DNA sequence that determines the development of limbs and other body parts in a fetus

HTML	Hypertext markup language
HUGE	Human Unidentified Gene-Encoded (Japan) database
HUNT	Human Novel Transcripts (Japan) database at the Helix Institute
IAB	Keio University's Institute for Advanced Biosciences (Japan)
ICSB	International Conference on Systems Biology
In silico	A trial carried out on a computer chip
In vitro	In an artificial environment outside the living organism
In vivo	Within a living organism
IP	Intellectual property
IPCR	Interdisciplinary Program in Cellular Regulation at the University of Warwick (U.K.)
IPTG	Isopropyl beta-D-thiogalactopyranoside
ISB	Institute for Systems Biology (U.S.)
ITB	Institute for Theoretical Biology (Germany)
IWR	Interdisciplinary Center for Scientific Computing (Germany)
JBIC	Japan Biological Informatics Consortium (Japan)
JBIRC	Japan Biological Information Research Center (Japan)
JSPS	Japan Society for the Promotion of Science (Japan)
JST	Japan Science and Technology (Japan)
KaPPA-VIEW	Kazusa Plant Pathway Viewer (Japan)
KDRI	Kazusa DNA Research Institute (Japan)
KEGG	Kyoto Encyclopedia of Genes and Genomes (Japan)
KGML	KEGG Markup Language (Japan)
LC-FT-MS	Liquid chromatography Fourier transformation mass spectrometry

LC-PDA-MS	Liquid chromatography photo-diode-array-detection mass spectrometry
LC-SPE-NMR-MS	Liquid chromatography, solid phase extraction, nuclear magnetic resonance and mass spectrometer
LC-TOF-MS	Liquid chromatography time-of-flight mass spectrometry
LSBM	Laboratory for Systems Biology and Medicine (Japan) at the University of Tokyo
LSI	Life Sciences Interphase (U.K.)
MALDI-TOF	Matrix assisted laser desorption/ionization-time of flight
MAPS	LIPID Metabolites and Pathways Strategy
MAPK	Mitogen-activated protein kinase
MB	Megabytes
MEK	A dual-specificity kinase that phosphorylates the tyrosine and threonine residues on ERKs 1 and 2 required for activation
MCA	Metabolic Control Analysis
MEGD	Microarray gene expression data
MEK	Mitogen-activated protein-ERK kinase
METI	Ministry of Economy, Trade and Industry (Japan)
MEXT	Ministry of Education, Culture, Sports, Science and Technology (Japan)
MGC	Mammalian Gene Collection at the National Institutes of Health (U.S.)
MHC molecule	Major histocompatibility complex molecule
MIPS	Munich Information Center for Protein Sequences (Germany)
MIT	Massachusetts Institute of Technology (U.S.)
MOAC	Molecular Organization and Assembly in Cells (U.K.)
MPI	Max Planck Institutes (Germany)

MRC	Medical Research Council (U.K.)
mRNA	Messenger ribonucleic acid
MS	Mass spectrometer/spectrometry
MW	Molecular weight
NADH	Nicotinamide adenine dinucleotide
NAIST	Nara Institute of Science and Technology (Japan)
NCBI	National Center for Biotechnology Information (U.S.)
NEDO	New Energy and Industrial Technology Development Organization (Japan)
NIGMS	National Institute of General Medical Sciences (U.S.)
NIH	National Institutes of Health (U.S.)
NMR	Nuclear magnetic resonance
NSF	National Science Foundation (U.S.)
NOW	Netherlands Organization for Scientific Research (Netherlands)
NP-hard	The complexity class of decision problems that are intrinsically harder than those that can be solved by a nondeterministic Turing machine in polynomial time.
OBIGRID	Open Bioinformatics Grid (Japan) at the RIKEN Yokohama Institute
ODE	Ordinary differential equation
ORFs	Open reading frames
P13K	Phosphoinositide 3-kinase
PCA	Principal component analysis
PCH	Polymerase chain reaction
PCR	Protein-coupled receptors
PDA-MS	Photo-diode-array-detection mass spectrometry
PDE	Partial differential equation

PDGF	Platelet-derived growth factor
PDM	Plant diagnostic module
PEST sequences	A sequence that has been associated with rapidly degraded proteins. The short life-time of a protein is signaled by a region rich in the amino acids proline (P); glutamic acid (E); serine (S); or threonine (T).
PIs	Principal investigators
PLoS	Public Library of Science Journal
PSC	Plant Science Center (Japan) at the RIKEN Yokohama Institute
QA/QC	Quality assurance/quality control
QUASI	Quantifying Signal Transduction (Europe)
RAFL	RIKEN *Arabidopsis* Full-Length (Japan)
RARGE	RIKEN *Arabidopsis* Genome Encyclopedia (Japan)
RCAI	Research Center for Allergy and Immunology (Japan) at the RIKEN Yokohama Institute
RCAST	Research Center for Advanced Science and Technology (Japan) at the University of Tokyo
RecSiP	Reconfigurable cellular simulation platform (Japan)
RIKEN	Institute of Physical and Chemical Research (Japan)
RISA	RIKEN Integrated Sequence Analysis (Japan)
RNAi	Ribonucleic acid interference
RSGI	RIKEN Structural Genomics/Proteomics Initiative (Japan)
RT-PCR	Reverse transcription-polymerase chain reaction
RTD	Research and Technological Development for the European Commission
RKT	Receptor tyrosine kinase
RRE	Rev-response element

SA node	Sino-atrial node
SAM	Shoot apical meristem
SAP	Systematic analysis of promoters (U.K.)
SBML	Systems biology markup language
SBW	Systems biology workbench
SCB	Society for Conservation Biology
SCSL	Sony Computer Science Laboratories, Inc (Japan)
siRNA	Small interfering ribonucleic acid
SNA	Stoichiometric network analysis
SNPs	Single nucleotide polymorphisms
SRC	SNP Research Center (Japan) at the RIKEN Yokohama Institute
SORST	Solution Oriented Research and Technology Program (Japan) at the Ministry of Education
SSA	Specific Support Activities (Europe)
SSP	Symbiotic Systems Project (Japan) at Keio University
STAT	Signal transducer and activator of transcription
STREPS	Specific Targeted Research Projects (Europe)
SysMO	Systems Biology of MicroOrganisms
SVG	Scalable vector graphics
TE	Tracheary elements
TEMBLOR	Concentrates on research and development to build the European Bioinformatics resources for the genomic era and beyond
Tflop/s	10^{12} floating point operations per second
TILLING	Targeting induced local lesions in genomes
TGFb	Transforming growth factor beta

TNF	Tumor necrosis factor
TNO	Institute of Environmental and Energy Technology (Netherlands)
TOF-MS	Time-of-flight mass spectrometry
tRNA	Transfer ribonucleic acid
UCL	University College London (U.K.)
ULB	Université Libre de Bruxelles (Free University of Brussels) (Belgium)
UCSB	University of California, Santa Barbara (U.S.)
UCSD	University of California, San Diego (U.S.)
UV	Ultraviolet
VCs	Venture capital firms
YAGNS	Yet Another Gene Network Simulator (Japan)